The Latin American Studies Book Series

Series Editors

Eustógio W. Correia Dantas, Departamento de Geografia, Centro de Ciências, Universidade Federal do Ceará, Fortaleza, Ceará, Brazil

Jorge Rabassa, Laboratorio de Geomorfología y Cuaternario, CADIC-CONICET, Ushuaia, Tierra del Fuego, Argentina

Andrew Sluyter, Louisiana State University, Baton Rouge, LA, USA

The Latin American Studies Book Series promotes quality scientific research focusing on Latin American countries. The series accepts disciplinary and interdisciplinary titles related to geographical, environmental, cultural, economic, political and urban research dedicated to Latin America. The series publishes comprehensive monographs, edited volumes and textbooks refereed by a region or country expert specialized in Latin American studies.

The series aims to raise the profile of Latin American studies, showcasing important works developed focusing on the region. It is aimed at researchers, students, and everyone interested in Latin American topics.

Submit a proposal: Proposals for the series will be considered by the Series Advisory Board. A book proposal form can be obtained from the Publisher, Juliana Pitanguy (juliana.pitanguy@springer.com).

More information about this series at http://www.springer.com/series/15104

Clara María Minaverry · Sebastián Valverde
Editors

Ecosystem and Cultural Services

Environmental, Legal and Social Perspectives in Argentina

 Springer

Editors
Clara María Minaverry
Consejo Nacional de Investigaciones
Científicas y Técnicas (CONICET), Instituto
de Ecología y Desarrollo Sustentable
(INEDES)
Universidad Nacional de Luján
Buenos Aires, Argentina

Sebastián Valverde
Consejo Nacional de Investigaciones
Científicas y Técnicas (CONICET), Facultad
de Filosofía y Letras
Universidad de Buenos Aires
Buenos Aires, Argentina

ISSN 2366-3421   ISSN 2366-343X   (electronic)
The Latin American Studies Book Series
ISBN 978-3-030-78380-8   ISBN 978-3-030-78378-5   (eBook)
https://doi.org/10.1007/978-3-030-78378-5

This Springer imprint is published by the registered company Springer Nature Switzerland AG
The registered company address is: Gewerbestrasse 11, 6330 Cham, Switzerland

To our families who always supported our career as scientific researchers and to all the scientists and professors who devoted their time to teach us in the past.

Foreword

The paradigm shifts that have been taking place in the Social Sciences in Argentina since the 1970s and 1980s are related to the politicization and organization of the indigenous and peasant movements in Argentina and Latin America.

This book, which is an excellent compilation by Clara Minaverry and Sebastián Valverde, is, to my judgment, the reflection of these paradigm shifts, and a valuable contribution to the debates in these times of pandemic and uncertainty regarding the future.

In the seventies, and in the face of an increasing demand for commodities, industrialization, and the growth of multinational corporations, indigenous peoples began to organize themselves politically to defend their territories, culture, and way of life.

The need to safeguard the territories traditionally inhabited by more than 800 indigenous peoples from the so-called *Latin America* forced communities to organize politically outside of their traditionally inhabited territories, contacting peasant movements and political parties (Nilo Cayuqueo, CIPIAL 2018).

Historically, the peasant movement has its roots in the Mexican Revolution of 1910. Its continuation was the Peasant Revolution in Bolivia in 1952 and in Peru in 1970. Despite the fact that most of these organizations were integrated by indigenous communities, the leadership of the organizations during this time was heavily influenced by left-wing parties, which presented class struggle as the tool and a socialist revolution as an alternative.

The social sciences, from a classical Marxist perspective, and with a certain degree of dogmatism, analyzed indigenous peoples as mainly peasant societies, although they recognized the need to preserve their cultures and traditions. The international participation of indigenous organizations, and the resulting enactment of national laws and international treaties, also influenced, without a doubt, to change the point of view of social sciences with regard to the way the topic of indigenous peoples was approached.

The historical meeting of Barbados in 1971 was the place where the anthropologists who studied the topic of indigenous peoples got together and invited indigenous leaders to debate. After the First International Conference of the

Americas that took place in Canada in 1975, which was organized by the Union of Indian Chief and where the World Council of Indigenous Peoples was establablished and also after the First Conference of Indigenous Peoples of the UN in Geneva in 1997, several important debates were produced among indigenous peoples and social scientists.

In the context of the fifth centennial of the conquest of the American continent, during the United Nations Conference on Environment and Development held in Rio de Janeiro in 1992, indigenous peoples demanded the need to protect the environment and all life on the planet, warning about the dangers of carrying out ecocide, the product of the extractivist exploitation of the environment. The so-called environment became the focal point of the struggles for a fairer and healthier world.

In Argentina, many have been the struggles to get the State to recognize us as legal subjects, and, without a doubt, the ratification of the ILO-Convention 169 in 1992 by the State has been one of the most important legal instruments in this recognition. However, this convention, like other international agreements and laws, is not applied and is breached with impunity. The bond between big capitalist ventures and the state regulatory agencies allows the continued violation, with impunity, of laws and international treaties, without consulting us, indigenous peoples, nor requesting our participation nor our free, informed, and prior consent, as is required by the ILO-Convention 169.

The territorial and environmental aspects are far from being solved, and this is exposed by the case of the Law 26,160, Bill of Indigenous Territorial Emergency of 2006, which prohibits forced evictions in territories where there is an ongoing conflict. Today, 14 years after the bill was passed, it had to be extended for a third time in 2017, without having, to this date, fulfilled its objective.

Conflicts grow at the pace extractive, real-estate and touristic activities do. The advance of national and international capital with public and private ventures and the lack of recognition of communal property of the territories increase the conflicts.

With the approval, in 1996, of the use of transgenic soybeans and its technological package, Argentina has consolidated, in the last two decades, agribusiness as its productive model for the export of commodities of forage to raise animals.

This productive model, which was established as public policy, has since derived in the increase of 848% in the use of poison in an area that was extended by 50%. Nowadays, 300 million liters of agrotoxins are being spilled, also in the production of food. Food is merely considered a commodity.

Biotechnology's impact has sustained the encroachment of the agricultural and livestock frontier from the *Pampas* region toward the Northern provinces, where the largest population of indigenous peoples and peasants are to be found.

The agroindustrial model, with the resulting expulsion of indigenous peoples and peasants toward the big cities, deforestation, loss of native forests, degradation of the soil, air and water pollution, has been an attack on our bodies and our territories.

Production is progress. Nature preservation is backwardness. The once called "deserted territory" that justified military campaigns is currently owned by the descendants of those military officers, big landowners, and under the control of the soy-agribusiness, vast areas sown with transgenics. These matrices of extractivist development are shared by right-wing neoliberal governments and those labeled as progressive.

The colonial states, both on the Argentine as on the Chilean side, conceived indigenous territories as lands to be exploited. Therefore, they promoted a discourse of civilization and barbarism, and the stigmatization of indigenous peoples through education, which brought, as a consequence, stark and violent racism, which national States and authorities, to this day, try to sustain. Currently, these policies have become ethnocidal, due to the devastation that they have produced in our territories, preventing the intergenerational transmission of our culture, both oral and written.

In the chapter written by Juan Carlos Radovich together with Sasha Camila Cherñavsky, Ailén Flores, Frank James Hopwood, Rocío Míguez Palacio, Nadia Molek, and Sebastián Valverde about the prejudice of the foreignness of the Mapuche people, one by one, the fallacies that compose this prejudice are analyzed. Its origin lies in the justification of the invasion and subsequent military and landowner occupation of the indigenous territories in the *Pampas* and Patagonia, and the subjugation of the indigenous peoples. In this way, the emerging Argentine and Chilean States consolidated their borders with the bloody invasions of these territories.

To push the indigenous peoples out of their territories, it was conjured up that the Mapuche had invaded Tehuelche territory, who are usually branded as the "real Argentine Indians". But as it is pointed out in this chapter, this accusation is fallacious, as there is overwhelming evidence that the Mapuche people lived on both sides of what is, nowadays, the Argentine-Chilean international border. The Andes mountain range, with its passages, is not a border for the indigenous peoples, who are additionally linked by kinship and cultural exchanges among the different territories. The Tehuelche and the Mapuche are peoples with a fraternal bond and have inhabited both sides of the Andes for thousands of years, and pre-exist both the Argentine and the Chilean States, as is recognized in the Argentine constitution amended in 1994.

Prejudice always implies ignorance, in many cases willful ignorance. It is well known that after prolonged diaspora, many peoples were able to reorganize themselves and reestablish their communal territorial logic in other places, which, nonetheless, continued to be besieged. This happened at first with the creation of big cattle farms whose owners had the support of repressive forces, with the purpose of dispossessing again indigenous people. Also, the aim was to proletarize them in order to turn them into cheap labor, and then expel them by the implementation of extractivist and tourism industries. The only possibility offered to indigenous peoples was either assimilation or extinction.

Any shred of autonomy that we achieve is judged as an attempt to establish a foreign nation within the Argentine territory and not as the right to self-government and to the communal territories that should be ours by right. As the chapter points out, the territorial emergency law and the surveys carried out within the context of this law, still incomplete for many communities, is a guarantee for the use of our territories. But we need a communal property law that recognizes the territories of indigenous peoples in Argentina.

"The best way to protect the forest is to protect the people who live in it," stated a technician of one of the projects for the protection of forests in the Chaco region, quoted by Natalia Castelnuovo in her article in this book, an enlightening statement.

The forests, from a non-extractivist perspective, are "communal" territories; they are an element of cultural identity for the peoples that live from and with them.

Forests, indigenous peoples and farmers are constantly under siege by agents of a system where the "communal," equality, and the respect for nature and its cycles is deliberately ignored in pursuit of the compulsive enrichment of a few. In this way, scrubland clearing, deforestation, and wildfires proliferate, to either expand the agribusiness farmland bound to exploitation, such as soybeans, as well as real-estate speculation that generates social segregation, at the expense of the peoples, biodiversity, and food sovereignty.

There are two competing paradigms of how to construct a society, to live life, that oppose each other: One includes and protects, and the other excludes and kills. States and governments, in a contradictory manner, protect both, with laws that benefit one another. The terminology used in these laws is an expression of this contradiction: "emergency," "minimal budget," "land-use planning," "conservation value." The range of hues of the traffic light, from the urban context, is used to designate whether forests are exploitable. Red, absolute prohibition, protects, and green, total liberation, enables destruction.

As Mariana Schmidt points out in her chapter, in the capitalist system, nature is associated with "backwardness," only admissible in either touristic or zoological terms, within "reservations," in "installments." Meanwhile, the logic of "production" associated with deceitful "progress" reigns: bread for today and just for some, hunger for tomorrow. Literally bread, the distribution of sacks of flour, and not the use of the forest with what it implies in terms of community and identity.

The indigenous–peasant reemergence occurs in the context of a social and environmental emergency, besieged by multiple fronts that threaten their own existence.

Therefore, the tools that the State provides must be functional to the strengthening of autonomy and consciousness, both collective and communal.

Capitalism drives land-grabbing practices directed toward obtaining profit and producing proletarians, cheap labor. It happened this way during the European "primitive accumulation" that expanded to other parts of the globe with colonialism and imperialism, and it is not stopping. But to indigenous people and peasants, forests are not just lands, these are territories of communal use, and it is therefore important that a law of communal property regulates what has been established in Article 75, Section 17, of the 1994 Constitution, where it recognizes "the communal possession and property of lands traditionally occupied." And also, Article 41, where it recognizes that all inhabitants "enjoy the right to a healthy, balanced environment, suitable for human development and so that productive activities meet current needs without compromising that of future generations; and have the duty to preserve it". Thereupon, this constitutional article assigns the authorities the duty to protect "this right, of rationally using natural resources, of preserving the natural and cultural patrimony and biodiversity, and of access to environmental education and information." But the issue is to define what "rational use" means. Additionally, already contained in the term "patrimony," a verticality and asymmetry is established regarding who determines what it actually entails, usually bureaucratic institutions, both national and international, the consequence of which is predatory tourism and rather than communal. We state that we are for "fratrimony," for new paradigms, for new forms of expression, for communal control, and for horizontal fraternity. As Clara Minaverry and Sebastián Valverde pointed out in their chapter, the right to biodiversity must be articulated with the right to cultural and social diversity, but autonomous and empowered, not mediated. True respect for nature implies the acknowledgment of the peoples that respect it, and of their ways of life, different from the predominant extractivist paradigm.

The case of the Lof Paichil Antrio and of the Las Huaytekas community, in the province of Río Negro, in Argentina, and the actions deployed around the restoration process is another aspect dealt with in the study of Valeria Inigo Carrera, Alejandro Balazote, and Gabriel Stecher. The Administración de Parques Nacionales [National Parks Administration] founded in 1922 took over the territories of indigenous peoples for the purpose of "preserving them in order to take care of the environment". However, parts of these territories have often been given to private companies for touristic exploitation and denying that right to the communities that live there.

We, the Mapuche, as most indigenous peoples, state that the earth does not belong to us, but that we belong to the earth. Therefore, it is our duty to protect and take care of it, so that the coming generations can have a more harmonious life. Land is not a resource. Land is forests and nourishment, scrubland and medicine, fuel and culture. In our worldview, the territory is a prerequisite for life, and the indigenous peoples and peasants are their main custodians.

This excellent compilation which includes comprehensive studies that help us to analyze critically the naturalization of actions and omissions by a State that, in the midst of the twenty-first century, is still presented as unbiased, but that, in times of globalization, mainly responds to privately owned capital and not to the peoples of this earth.

I congratulate Clara Minaverry and Sebastián Valverde, and all the authors that have worked on this book, that without a doubt will not only help to inform the public, but also start an international debate about the future of this planet, that is not only ours, us human beings, but of all lives.

Buenos Aires, Argentina   Nilo Cayuqueo
October 2020

Nilo Cayuqueo was born and grew up in the Lof called "Campo de la Tribu de Coliqueo," in the province of Buenos Aires (300 kilometers west of Argentina's capital). He is a member of La Azotea Mapuche Community and of the Mesa de Pueblos Originarios of the province of Buenos Aires. He has established several indigenous organizations and has participated in several international meetings. He was forced to flee Argentina due to the political persecution in those times and has lived in Europe and the USA, where he always participated in several international indigenous organizations and movements.

Preface

This book is one of the products of a multidisciplinary research project entitled "Cultural Ecosystem Services in the Indigenous Communities of the Provinces of Río Negro and Neuquén in Argentina: Implementation of Public Polices and Application of the National Law of Native Forests" presented at the Universidad Nacional de Luján in Argentina which was developed from 2019 until 2021.

The main objective of this research project was to analyze solutions by the elaboration and design of interdisciplinary public policies in order to comply with the National Forest Law 26,331 of Argentina, which was enacted in 2007.

Another purpose of this project was to identify the different impacts produced by the application of forests regulations to the indigenous people and small producers of the area of the Nahuel Huapi National Park area and surroundings, in the provinces of Río Negro and Neuquén in Northern Patagonia in Argentina. September, 2021

Buenos Aires, Argentina

Clara María Minaverry
Sebastián Valverde

Acknowledgments

To the Consejo Nacional de Investigaciones Científicas y Técnicas (CONICET) (National Scientific and Technical Research Council) and Universidad Nacional de Luján, in Argentina.

We appreciate the careful reading of this book by Pamela Pulcinella, editor of Facultad de Filosofía y Letras, Universidad de Buenos Aires, Argentina.

About This Book

This book describes different cases and experiences in connection with environmental and cultural ecosystem services in Argentina. It contributes to the debate in connection with different approaches to analyze ecosystem services, focusing on the environment, the law and several social perspectives and concerns. Among the topics discussed are the implementation of the National Forests Act, the regulation of ecosystem services, the role of indigenous people, and the policies in place for nature and forests conservation in the different regions of Argentina.

This publication describes the contributions provided by cultural ecosystem services (especially from native forests), and it analyzes if they are adjusted, incorporated, or protected in public policies and/or regulations. In this regard, there is a scarce regulatory and jurisprudential development in connection with cultural ecosystem services in Argentina and in Latin America, which mainly affects the quality of life of indigenous and rural communities living in and around native forests. They are the key actors who will be deeply examined in the present publication, and this issue can be considered the major strength.

This book is one of the few research studies about cultural ecosystem services developed in Latin America and presents an attractive combination of the legal, environmental, and social approaches and was written by an interdisciplinary team of academics who have theoretical and practical experience in this region where there is a valuable natural and cultural heritage.

The display of the book is the following one:

Chapter 1 is "General Remarks About the Environmental Law and Anthropology and the Impacts Among Forests, Communities, and Regulations" by Clara María Minaverry and Sebastián Valverde. It is divided in two parts. The first one will provide general information about deforestation and indigenous and small producer populations, and the second one will focus on Environmental Law in Argentina which relates to ecosystem and cultural services provided by forests.

Chapter 2 "Environmental Policies and Territorial Conflicts in Argentina—From the Deforestation of Native Forests to Agrochemical Spraying in the Province of Salta" by Mariana Schmidt will try to unravel the and challenges ahead toward an environmental, social, and territorial scenario in the context of the implementation of the Argentine National Act 26,333 of "National Forests," enacted in 2007. It is focused on the case of the province of Salta, in Northern Argentina.

Chapter 3 "Regulatory Progress and Social Perception of the Ecosystem Services Provided by Urban Forests in the Municipality of Luján, Province of Buenos Aires, Argentina" by Analía Scarselletta and Elena Beatriz Craig, will analyze the local regulatory framework for the recognition, assessment, and protection of the services provided by urban forests, and it explores the social perception about them. This is a qualitative investigation of an exploratory, documentary, and descriptive nature.

Chapter 4 "Introduction—Indigenous Presence and Current Legislation in Argentina" by Juan Manuel Engelman and Sofía Micaela Varisco will provide general information about the presence of indigenous peoples in Argentina, as well as the main laws toward them at the present.

Chapter 5 "Forests, Culture and Health" by Mariana Costaguta, Martín Rodríguez Morcelle, Laura Gabucci, and Bruno Lus will focus on describing the territory as a multicultural space with traditional and circumstantial and recent knowledge and practices, which presents a complexity and dynamics which enriches itself. Also, biological and cultural diversity nourishes with migrant population exchanges, which are generations that provide benefits with popular health construction.

Chapter 6 "Forests and Public Policies in the Argentine Northern Patagonia Region—Small Producers, Capitals, and Territorial Claims" by Valeria Iñigo Carrera, Alejandro Balazote, and Gabriel Stecher will examine the implications of public policy regarding nature conservation in general and territorial planning of native forests in particular, in relation to the territorialities configured by different social subjects, focusing specifically on small-scale agricultural producers (indigenous and non-indigenous). A supplementary aim is to problematize the dynamics of increasing territorial conflicts in the forest areas of the mountain range region of Northern Patagonia (in both provinces), sparked by the development of various public and private ventures on land and territories occupied by the aforementioned small-scale producers.

Chapter 7 "Indigenous and Peasants Lands Under the Spotlight—A State Forest Policy in the Gran Chaco, Argentina" by Natalia Castelnuovo Biraben will analyze national forestry policy aimed at peasants and indigenous peoples who inhabit the Gran Chaco area, in Argentina, whose lands and life systems have been affected by deforestation, the expansion of the livestock farming model and what some researchers call "world land fever."

Annex I is *Notes on the Academic Report on the Impact of COVID-19 on Indigenous Peoples*, by Sebastián Valverde. This is a summary of the main parts of the report presented in July 10, 2020, by the Observatorio Universitario de Buenos Aires (OUBA) entitled "Quinientos años no es nada" ("Five thousand years is nothing"), which is related to the report about the socio-economic and cultural effects of COVID-19 and of social, preventive, and compulsory isolation in the indigenous peoples of Argentina.

Annex II is *Prejudice Towards Mapuche People: The Attribution of Foreignness as a Strategy for Stigmatization*, by Juan Carlos Radovich, Sasha Camila Cherñavsky, Nadia Molek, Rocío Miguez Palacio, Ailén Flores, Frank James Hopwood, and Sebastián Valverde. It describes the main seven prejudices against these communities.

Contents

Editors and Contributors

About the Editors

Clara María Minaverry completed her Ph.D. in Law, School of Law, University of Buenos Aires, Argentina (2014). She received her Master's Degree in Environmental Law, Universidad Complutense de Madrid, Spain (2009), and she is a Lawyer, School of Law, Universidad de Buenos Aires, Argentina (2002).

She is Permanent Researcher specialized in Environmental and International Environmental Law at the Consejo Nacional de Investigaciones Científicas y Técnicas (CONICET) in Argentina and at the Institute of Ecology and Sustainable Development, Universidad Nacional de Luján.

She is Professor of Environmental Law at the Social Sciences Department at the Universidad Nacional de Luján and a director of several scientific projects related to legal forests, water, and ecosystem services protection.

She is also Postgraduate Professor at the Universidad de Buenos Aires and at the Universidad Tecnológica Nacional, in Argentina and acted as consultant of international agencies.

Sebastián Valverde completed his Ph.D. in Social Anthropology, Facultad de Filosofía y Letras, Universidad de Buenos Aires, Argentina (2006).

He has a degree (2001), and he is Professor (1998) in Anthropological Sciences at the same school. He is Permanent Researcher at the Consejo Nacional de Investigaciones Científicas y Técnicas (CONICET), at the Facultad de Filosofía y Letras, Universidad de Buenos Aires, where he is also Professor at the Department of Anthropological Sciences.

He also teaches, as regular Professor at the Social Sciences Department at the Universidad Nacional de Luján. He integrates and directs different research projects in both universities, related to economic anthropology, interethnic relations, and ethnopolitical movements of indigenous peoples related to the Mapuche indigenous people settled in southern Argentina. He has published articles in specialized magazines and books. He has also directed and supervises different postgraduate thesis and undergraduate scholarships for students (UBA) and postgraduate (CONICET, FONCYT, and UBA).

Contributors

Alejandro Balazote Universidad de Buenos Aires (UBA), Buenos Aires, Argentina;
Social Sciences Department, Universidad Nacional de Luján (UNLu), Luján, Argentina

Natalia Castelnuovo Biraben Consejo Nacional de Investigaciones Científicas y Técnicas (CONICET), Facultad de Filosofía y Letras, Universidad de Buenos Aires, Buenos Aires, Argentina

Mariana Costaguta Vice Coordinator, TRAMIL, Buenos Aires, Argentina

Elena Beatriz Craig Technology Department, Universidad Nacional de Luján, Luján, Argentina

Juan Manuel Engelman Consejo Nacional de Investigaciones Científicas y Tecnológicas Facultad de Filosofía y Letras, Universidad de Buenos Aires and Social Sciences Department, Universidad Nacional de Luján, Argentina, Buenos Aires, Argentina

Laura Gabucci Universidad Nacional de Luján, Luján, Argentina

Valeria Iñigo Carrera Universidad Nacional de Río Negro (UNRN) – Consejo Nacional de Investigaciones Científicas y Técnicas (CONICET), Instituto de Investigaciones en Diversidad Cultural y Procesos de Cambio (IIDyPCa), San Carlos de Bariloche, Argentina

Bruno Lus Universidad Nacional de Luján, Social Sciences Department, Luján, Argentina

Clara María Minaverry Consejo Nacional de Investigaciones Científicas y Técnicas (CONICET), Buenos Aires, Argentina;
Institute of Ecology and Sustainable Development-INEDES, Universidad Nacional de Luján, Luján, Argentina

Martín Rodríguez Morcelle Universidad Nacional de Luján, Luján, Argentina

Analía Scarselletta Social Sciences and Technology Departments, Universidad Nacional de Luján, Luján, Argentina

Mariana Schmidt Consejo Nacional de Investigaciones Científicas y Técnicas (CONICET), Facultad de Ciencias Sociales, Universidad de Buenos Aires, Buenos Aires, Argentina

Gabriel Stecher Universidad Nacional del Comahue (UNComa), Asentamiento Universitario San Martín de Los Andes, Cátedra Extensión Rural, San Martín de los Andes, Argentina

Sebastián Valverde Consejo Nacional de Investigaciones Científicas y Técnicas (CONICET), Social Sciences Department Universidad Nacional de Luján, Luján, Argentina;
Consejo Nacional de Investigaciones Científicas y Técnicas (CONICET), Facultad de Filosofía y Letras, Universidad de Buenos Aires, Buenos Aires, Argentina

Sofía Micaela Varisco Fondo para la Investigación Científica y Tecnológica, Programa de Arqueología Histórica y Estudios Pluridisciplinarios, Social Sciences Department, Universidad Nacional de Luján, Luján, Argentina

List of Figures

**Forests and Public Policies in the Argentine Northern Patagonia
Region: Small Producers, Capitals, and Territorial Claims**

Annexes

List of Tables

General Remarks About the Environmental Law and Anthropology and the Impacts Among Forests, Communities, and Regulations

Regulatory Progress and Social Perception of the Ecosystem Services Provided by Urban Forests in the Municipality of Luján, Province of Buenos Aires, Argentina

Forests and Public Policies in the Argentine Northern Patagonia Region: Small Producers, Capitals, and Territorial Claims

General Remarks About the Environmental Law and Anthropology and the Impacts Among Forests, Communities, and Regulations

Clara María Minaverry and Sebastián Valverde

Abstract Most of the indigenous peoples in Argentina are urban and the minority are small producers. On the other hand, in several cases, these last ones do not belong to an indigenous category, so only a minority belong to both categories simultaneously. Also, this composition changes radically depending on the different regions of the country and placing them away from their natural area: the forests. In this context, cultural ecosystem services provided by forests can be defined as those connected with provision and regulation functions, spiritual and educational dimensions, which are a part of a cultural function of ecosystems. In particular, we discovered that in most cases, these kinds of services are not being recognized, protected nor regulated by case law or actual regulations, except in some countries of the region where it is starting to change this situation. Also, there are several setbacks which relate to environmental law ineffectiveness regarding the implementation of valid regulation of native forests protection and of ecosystems services.

Keywords Environmental law · Indigenous communities · Forests

C. M. Minaverry (✉)
Consejo Nacional de Investigaciones Científicas y Técnicas (CONICET), Luján, Argentina
e-mail: cminaverry@unlu.edu.ar

C. M. Minaverry
Institute of Ecology and Sustainable Development - INEDES, Social Sciences Department, Universidad Nacional de Luján, Luján, Argentina

S. Valverde
Consejo Nacional de Investigaciones Científicas y Técnicas, Social Sciences Department, Universidad Nacional de Luján, Luján, Argentina

S. Valverde
Consejo Nacional de Investigaciones Científicas y Técnicas (CONICET), Faculty of Philosophy and Letters, Universidad de Buenos Aires, Buenos Aires, Argentina

© The Author(s), under exclusive license to Springer Nature Switzerland AG 2021
C. M. Minaverry and S. Valverde (eds.), *Ecosystem and Cultural Services*, The Latin American Studies Book Series, https://doi.org/10.1007/978-3-030-78378-5_1

1

1 Introduction to Social Anthropology in Argentina

1.1 Agriculturization, Soyzation, and Deforestation Processes in Argentina

A process of "agriculturization" or "*soyzation*" is taking place in Argentina which produced an expansion of the agricultural frontier with a deep transformation of the rural area and of social relationships, especially in the decade of the 1990s which was increased in the last years. These changes are related to the biotechnological progress, the dynamization of real estate, and the increasing foreignization and concentration of lands and of agricultural exports (Schmidt 2017). Meanwhile, a process named "*bovinization*" occurred and implied a relocation of cattle providing from the Humid Pampa toward other "peripheral" regions such as the Northeast and Northwest of Argentina and other regions outside the *Pampas* region (Murphy and Grosso 2012; Paz 2015).

These dynamics implied the raising of territories used for these activities replacing areas without cultivation or considered marginal in terms of capital or replacing other crops with annual rotations such as the case of soybean (Paruelo et al. 2006).

The main consequences of this process are related to the raising of environmental degradation, erosion and salinization of soils, biodiversity, contamination of soils and water, health problems connected to the exposure to agrochemicals, land concentration and expulsion of population to the peripheral urban areas (Schmidt 2017). Also, we must take into consideration that agrochemicals application affects the health of the population (Naharro and Álvarez 2011).

It also relates to problems connected with climate change (floods, desertification, and other) which radically impacts on the quality of life of indigenous and small rural producers, on the socio-communitarian dynamics and in the interaction with the State, private actors, NGO's, and others (Castilla 2017). These transformations determine new conditions of social exclusion which exacerbated the existing ones because they were expelled as workers (mainly as a consequence of the mechanization of crops) producing a progress in the dispossession of common goods such as lands and water (Paz and Fleitas 2019).

Small rural producers who practice agriculture as a subsistence and as producers of food for local and regional markets are being negatively affected by this productive model implemented by concentration, privatization of lands, and over exploitation of natural resources (Paz and Fleitas 2019). In connection with this, several disputes, mobilizations, and demands filed to public authorities with multiple evictions and episodes of territorial violence started to occur (Aguiar et al. 2018; Castilla 2018; Schmidt 2017).

Other of the main issues analyzed here in this book is the deforestation, in particular of shrubland areas and of native forests. Argentina in the last decades of the twentieth century and the beginning of the twenty-first century has experimented a sad record being one of the countries which had higher levels of deforestation of native forests in the world. It had 35 million of hectares in 1987, and it was reduced to 31 million in 1998 and afterward descended even more to 27 million (National Environmental and Sustainable Development Ministry 2018).

In this regard, deforestation continued without complying with National Law 26,331 enacted in 2007 entitled "Minimum Standards for Environmental Protection of Native Forests," known as the "Forests Law." Deforestation in the region of Chaco is the second peak in South America (after the Amazonas) replacing forests and shrublands through the expansion of the agro-industrial frontier. This situation represents alarming indexes about deforestation in some of the provinces in Northern Argentina, such as Santiago del Estero, Salta, Chaco, and Formosa (Table 1).[1]

Since several decades ago, the loss of shrublands and native forests have been producing several effects on the natural habitats and is one of the causes of emigration to the cities collaborating with the growing urbanization of indigenous communities and of small rural producers.

1.2 Small Producers and Indigenous Communities in Argentina

It is particularly important to define the meaning of "farmer" because there is an absence of this category in Argentina, and in its place the one which is used is "small producer." Scheinkerman de Obschatko et al. (2007) developed a report in the area of Familiar Agricultural Secretary (Scheinkerman de Obschatko et al. 2007: 21) which considers a small producer:

"(...) the heterogenous group of producers and their families (...) [who] participate in a direct way in the production with its physical work and productive management, not contracting permanent workers, having limitations with lands, capital and technology".

In Argentina, according to the Census of population of 2001 and the National Agricultural Census of 2002, there are 333,477 agricultural exploitations. These are mainly small producers taking into consideration the classification of the authors mentioned before, a total of 218,868 (66% of the total) (Scheinkerman de

---

[1]The provinces which were more affected by this process in absolute terms are Santiago del Estero, Salta, Chaco, and Formosa, which represent around 78% of the deforested territories between 2007 (when the Forests Law was enacted) and 2017 (National Environmental and Sustainable Development Ministry 2018).

Table 1 Surfaces of hectares of native forests in Argentina between 1937 and 2018

Year	1937	1987	1998	2002	2014
Surface of native forests (Ha)	37,535,308	35,180,000	31,443,873	30,073,385	26,428,349

Source National Agricultural Census (1937 and 1987), Data from the National Forest Institute (1998), Update from UMSEF—Forests Agency (2014), Environmental status Report (2018)—National Ministry of Environment and Sustainable Development of Argentina: https://www.argentina.gob.ar/sites/default/files/compiladoiea2018web.pdf

Obschatko et al. 2007).[2] In these numbers, we must include the family and the workers, so this figure rises to 775,296.[3]

How can we relate this category with the other fundamental one which must be defined in this book: indigenous peoples?

In Argentina and following the data provided by the last National Census of Population and Households of 2010, there are more than a million of individuals who belong to indigenous peoples,[4] from at least forty communities settled in different regions of the country.

The proportion of indigenous in Argentina, according to this source, is of 2.4% and the criteria to define it is the self-belonging (National Constitution and ILO Agreement 169).

Most of the indigenous peoples in Argentina are urban and the minority among them are the small producers. On the other hand, in several cases, these last ones do not belong to an indigenous category, so only a minority belong to both categories simultaneously. Also, this composition changes radically depending on the different regions of the country.

In Northeastern and Northwestern Argentina, due to the higher incidence of the agricultural activities in the regional economies, the presence of rural populations (indigenous and non-indigenous) exceeds, in some cases, the 30, 40 or 50%. In most of these regions, deforestation levels are extremely high due to the expansion of the agricultural frontier and the a*griculturization* and *sojización* (soybean expansion), producing severe consequences in small rural producers and/or in indigenous peoples.

---

[2]We must consider that this data is 13 years old. Notwithstanding, it is a valid source because it presents coherent information, and there are no other inputs (which functions as a discontinuity indicator and of the analysis of public policies).

[3]This desegregates in producers (or partners) in 340,735 peoples; families of the producer 204,457- and other people who is not family related to the producer: 229,690.

[4]The indigenous peoples of Argentina are the following: qom (toba), mbya guaraní, moqoit, mapuche, guaraní, tupí guaraní, avá guaraní, kolla, diaguita, diaguita-calchaquí, wichí, huarpe, quechua, aymara, nivaclé (chulupí), tonokote, omaguaca, tastil, günün a küna, comechingón, comechingón-camiare, ocloya, iogys, chané, tapiete, iyofwaja (chorote), sanavirón, ranquel, wehnayek, atacama, lule, quilmes, mapuche-pehuenches, tehuelches, mapuche-tehuelches, selk 'nam, haush y selk 'nam-haush.

Table 2 Northwestern provinces of Argentina—Evolution of total, rural and urban population—Comparison between 1960 (*) and 2010 (**)

Provinces	Year	Total population	Urban population	Rural population	Urban population (%)	Rural population (%)
Catamarca	1960	168,231	70,570	97,661	41.95	58.05
	2010	367,828	283,706	84,122	77.13	22.87
Jujuy	1960	241,465	118,665	122,800	49.14	50.86
	2010	673,307	588,570	84,737	87.41	12.59
La Rioja	1960	128,220	54,658	73,562	42.63	57.37
	2010	333,642	288,518	45,124	86.48	13.52
Salta	1960	412,854	226,899	185,955	54.96	45.04
	2010	1,214,441	1,057,951	156,490	87.11	12.89
Santiago del Estero	1960	476,503	167,944	308,559	35.25	64.75
	2010	874,006	600,429	273,577	68.70	31.30
Tucumán	1960	773,972	420,837	353,135	54.37	45.63
	2010	1,448,188	1,170,302	277,886	80.81	19.19

(*) *Source* 1960. National Census of Population—Volume I of Argentina. National Agency of Statistics and Census, Economy Secretary
http://www.estadistica.ec.gba.gov.ar/dpe/Estadistica/censos/030%20-%201960-Censo%20Nacional%20de%20Poblacion.%20Total%20Pais/PDF/1960.pdf
(**) *Source* 2010. Estimations elaborated using the National Census of Population and Households 2010 (INDEC), National Geographical Institute https://www.ign.gob.ar/NuestrasActividades/Geografia/DatosArgentina/Poblacion2

The following charts describe the comparison between rural, urban, and total population for each one of these departments in the provinces of Northeastern and Northwestern Argentina, opposing the year 1960 with the last census (2010). We can show how the rural area was depopulated during this century (Tables 2 and 3).

In these provinces, there are increasing conflicts with small producers, either *criollos* and indigenous. Some of them have been published in national media, such is the case of the deaths of children due to malnutrition in the wichi and guaraní indigenous communities during 2011 and 2015 (Naharro 2018).

Also, in the summer of 2020 before Covid-19, the situation was severe due to the effects provided by deforestation, loss of ancestral territories, soil contamination, lack of access to water and its consequences to health, malnutrition and the impossibility to fulfill basic human needs by the indigenous communities.

Table 3 Provinces of Northeastern Argentina—Evolution of the total rural and urban population
—Comparison between the years 1960 (*) and 2010 (**)

Provinces	Year	Total population	Urban population	Rural population	Urban population (%)	Rural population (%)
Corrientes	1960	533,201	247,312	285,889	46.4	53.6
	2010	992,595	822,224	170,371	82.8	17.2
Chaco	1960	543,331	252,139	291,192	46.4	53.6
	2010	1,055,259	892,688	162,571	84.6	15.4
Formosa	1960	178,526	72,166	106,360	40.4	59.6
	2010	530,162	428,703	101,459	80.9	19.1
Misiones	1960	361,440	115,096	246,344	31.8	68.2
	2010	1,101,593	812,554	289,039	73.8	26.2

(*) *Source* Census 1960
(**) *Source* 2010. Estimations elaborated using the National Census of Population and Households 2010 (INDEC), National Geographical Institute https://www.ign.gob.ar/NuestrasActividades/Geografia/DatosArgentina/Poblacion2

In this context, it was necessary to declare the socio-sanitary emergency in three departments in the province of Salta (Mancinelli 2019).[5] In these affected areas, we can see, according to the information provided by the census of 2001 and 2010, a growing urbanization of the population in particular the indigenous who lost their territories and original terrains (Flores Klarik 2019; Castelnuovo 2019).[6]

On the other side, in the case of the indigenous peoples, the environmental and sanitary risks related to other problem with a historical deepness: the access and tenure of lands (Castelnuovo 2019). The precariousness, the slow progress, and the delays in the execution of the National Law 26,160 of "indigenous territorial emergency" made this context more serious.[7]

In 2020, the Inter-American Court of Human Rights ruled in favor of the indigenous communities of Salta in the judicial cause filed in 1998 by the

---

[5]On January 29th, 2020, the socio-sanitary emergency was declared in the departments of San Martín, Rivadavia, and Orán in the province of Salta, as a consequence of the death of eight children of the indigenous communities due to gastrointestinal and respiratory complications which were accentuated by the acute malnutrition and dehydration (Resident Coordinator Office of United Nations 2020). At the moment, we are making final arrangements of this text (December 2020), a similar situation is happening and an indigenous wichi girl died because of the same cause. This situation is worse during the months of summer due to high temperatures (INFOBAE 2020), published on December 16, 2020.

[6]Flores Klarik (2019) instead analyzed the composition of urban and rural indigenous population in these districts from the analysis of the direct effects on the life of rural inhabitants, as a consequence of the expansion of the agricultural frontier in these departments. In all the cases, we can state that between 2001 and 2010, there was a relevant growth of the urban population in these departments.

[7]In Chap. 4 of Engelman and Varisco, they analyze in detail the difficulties of the implementation of this law which is related to the indigenous peoples and their connection with the "Forests Law".

Table 4 Departments of the province of Salta with higher levels of deforestation and with presence of indigenous population (where the socio-sanitary emergency was declared in January, 2020)—Evolution of urban and rural indigenous peoples between 2001 and 2010 (*) and deforestation of each department (**)

Departments—Province of Salta	Percentage of rural and urban indigenous population of the total population (indigenous and non-indigenous)—Comparison between				Deforestation (hectares)		
	2001		2010		Accumulated until the year 2000	Period 2001–2010	Period 2011–2018
	Rural (%)	Urban (%)	Rural (%)	Urban (%)			
Anta	0.9	1.1	0.9	1.1	376,559	361,036	156,100
San Martín	8.8	8.0	6.2	7.9	158,597	153,136	58,437
Orán	2.3	9.0	1.1	10.1	67,363	42,306	59,444
Rivadavia	35.1	3.3	27.5	6.1	16,928	83,678	49,773

(*) *Source* Author's elaboration according to the information provided by Flores Klarik (2019)
(**) *Source* Author's elaboration according to the information provided by the "Project of monitoring of deforestation in Dry Chaco," http://monitoreodesmonte.com.ar/

"Asociación de Comunidades Indígenas Lhaka Honhat" (Indigenous Communities Association Lhaka Honhat). The ruling set that the State should guarantee the regularization of the territory, cultural identity, healthy environment, food and water rights of the different indigenous communities of the Department of Rivadavia, in the Northeastern portion in the province (being this one of the poorest regions of Argentina) (Table 4).

The region of Patagonia (Southern Argentina) did not have an expansion of the agricultural frontier as it has happened in the Northern portion of the country as a consequence of the existence of much lower temperatures, the different characteristics of the soils, etc., and because it is not at all viable to plant crops such as soybean. Notwithstanding, in this region, there was an expansion of the touristic real estate frontier due to the investment in areas with landscapes with high value, which generates different effects on indigenous peoples and *criollos'* populations (as it will be described later in this book) (Valverde et al. 2021).

In this context, an increasing amount of indigenous peoples and small rural producers had to settle in different cities searching for work alternatives, mainly in poor neighborhoods creating new areas of migrants with precarious infrastructure, with overcrowding conditions, scarce resources to manage their domestic economy and with informal and temporary jobs (Castilla et al. 2019).

These tendencies were confirmed by the relevant reduction of agricultural exploitations in the census of 1988, 2002, and 2018 and were defined as "without boundaries" which represent the most precarious ones as it is shown in the following charts (Table 5).

Table 5 Evolution of the total number of agricultural operations—*Source* Agricultural Census

Farming establishments (Quantity)	National agricultural census 1988	National agricultural census 2002	National agricultural census 2018
With defined limits	378,357	297,425	228,375
Without defined limits	42,864	36,108	22,506
Total	421,221	333,353	250,881

The last Agricultural Census (2018) stated that from 2002 until 2018, 82,652 agricultural exploitations disappeared and also 103,000 more during the decade of the nineties. In this context, during the last 30 years, almost 200,000 agricultural exploitation vanished causing the loss of more than 900,000 jobs positions in the rural sector (Peretti 2020).

There is a rural depopulation, and these people were incorporated in different productive activities in urban areas. This explains that according to the data provided by the Census of 2010, most indigenous peoples settled in the regions of Buenos Aires and La Plata. They belong to indigenous communities, in the first place, from Northwestern Argentina and in the second, from the Northeastern area. Several authors (Tamagno and Maidana 2011; Maidana 2013; Engelman 2019) stated that in many cases in the first stage, there is a rural–urban migration from the same region and in the second, in some opportunities, there are extra-regional movements (from Northwestern and Northeastern regions to cities which have higher economic income such as Rosario, the "Great Buenos Aires" and La Plata).

2   Introduction to Environmental Law in Argentina

2.1   *Ecosystem and Cultural Services*

Ecosystem services appeared for the first time in the academic world after the publication of the Millennium Ecosystem Assessment and were defined as all the benefits that human populations obtain from ecosystems (MA 2005), which contributes to make life not only physically possible but also worth living (Daily 1997).

They are considered a useful tool because they causally link ecosystems to human needs (Balvanera et al. 2012: 9). In this sense, one of the main targets of this concept is to show that ecosystems in themselves create valuable services, which in most cases turn out to be more relevant than those achieved by its extraction and exploitation (Costanza et al. 2017).

Other authors have given a confusing connotation to the definition of Millennium Ecosystem Assessment, because it does not differentiate "ecosystems benefits" from the contributions made by those ecosystems, while the first require the use of capital (material and financial) and labor, and therefore, the outcome does not always translate into benefits to society. For that reason, the analysis of ecosystem services requires considering the way that different social agents may profit from them, which suggests the existence of an interrelation between the diverse biological, social, and cultural features of the environment. This may be complemented by the notion that ecosystem services pretend to analyze the different types of connections between society and nature or the way different social agents may profit from the services provided by ecosystems. This context suggests the analysis of the ecological, biological, social, and cultural features of the environment (Quétier et al. 2007; Ferro and Minaverry 2019).

Despite its controversial nature, the concept of ecosystem services has enhanced its institutionalization in the international agenda with the passing of the 17 Sustainable Development Goals in the framework of the United Nations Organization, which were developed by the end of the year 2015 and are estimated toward the year 2030. Some of them exclusively refer to the area of this book, for example, N° 15 states the obligation to: "Protect, restore and promote sustainable use of terrestrial ecosystems, sustainably manage forests, combat desertification, halt and reverse land degradation and biodiversity loss" (Minaverry 2020).

In this context, cultural services encompass representations and traditions which are associated with natural cycles or heritages (parties, rituals, sacred places, artistic representations) and also services linked to landscapes, local traditions, or services which collaborate with the creation of knowledge and to development of science (Costanza et al. 1997).

Cultural ecosystem services provided by forests can be defined as those connected with the provision and regulation functions and with spiritual and educational dimensions, which are a part of a cultural function of ecosystems. In particular, we discovered that in most cases, these kinds of services are not being recognized, protected nor regulated by case law or actual regulations, except in some countries in the region.

In this regard, climate change impacts depend on forests and its environmental services and is a problem which is constantly studied by scientists all around the world. There is an urgent need to start focusing on adaptation strategies and actions to promote the protection of ecosystem services and forests.

In Argentina, in December 2019, the National Law for adaptation and mitigation of global climate change was enacted and it is relevant for forests and ecosystem services protection. This problem is directly connected with the content of this book because this National Law states as one of its objectives to develop strategies, measures, policies and instruments related to the study of the impact, the vulnerability and the activities of adaptation to climate change which can provide development to humans and ecosystems.

2.2 The National Forests Law and the Regulation of Ecosystem Services in Argentina

States organize and function through different governmental systems which differ from each other and sometimes interfere with the application of the ecosystem approach. This form of State is structured through a group of entities (the provinces) which originally were separated without having any relationship between them and which have the faculty to enact its own legislation observing the limits set by the National Constitution. In this sense, the division of the administration of natural resources through provincial jurisdictions is complex, and it presents challenges in legal practice.

Argentina has adopted a federal system of government which includes 23 provinces, the city of Buenos Aires and the National Government. Pursuant to the Constitution, the provinces have the authority to pass their own legislation within their territorial limits.

The environmental paradigm was incorporated in 1994 in the amended National Constitution in section 41, third paragraph, whereby "The Nation shall regulate the minimum protection standards, and the provinces necessary to reinforce them, without altering their local jurisdictions." Furthermore, National Law 25,675 passed on 2002 (called "General Environmental Law"), and sets forth in section 6 that "the minimum protection standards, of section 41 of the Constitution, shall be understood as any law that grants environmental protection to the national territory as a whole, and has the purpose of establishing the necessary conditions to ensure environmental protection."

This implies that minimum protection standards laws about native forests must provide the main guidelines in order that provincial authorities could incorporate them when enacting their local laws.

However, the last paragraph of section 124 of the National Constitution becomes fundamental by stating that: "The provinces have the original dominion over the natural resources existing in their territory." Therefore, national and provincial levels act jointly, and each jurisdiction can adopt decisions to the detriment of an integral native forests management. This is extremely complex to apply in practice because it is not clear which is the limit of action of the national and provincial jurisdiction.

There is no current standing legislation dedicated to the protection of the environmental or ecosystem services, in any of the three analyzed jurisdictions (national, provincial, or municipal). However, some laws provide an indirect regulation.

One of them is the National Law 26,331 which was enacted in 2007 and was the first one for the protection of native forests in Argentina, and it was regulated by Decree 91 in 2009, more than a year after the enactment of the law.

The structure of the present legislation is of 44 articles and of one Annex. It is divided into 12 Chapters: 1. General Articles; 2. Territorial Planning; 3. Authorities; 4. National Program of Forest's protection; 5. Authorizations for land-use change or forest management; 6. Environmental impact assessment; 7. Hearings and Referendums; 8. Offender's Registry Office; 9. Inspections; 10. Sanctions; 11. National Fund for the enrichment and conservation of native forests; 12. Complementary Articles (Minaverry 2013).

The National Law 26,331 sets the minimum basis for native forests protection and ecosystem services, which can be exceeded by any province in order to enact more strict laws but could never be less demanding than the national ones. The aim of this law is to provide environmental protection standards, for the enrichment, restoration, conservation, and sustainable use of native forests and of their environmental services. Section 2 of the law states that this legislation also protects other natural resources not only being limited to forest but also to the whole ecosystem which surrounds them (Minaverry 2013).

This law was the first one in Argentina which recognized a payment mechanism for environmental services. Then, there is the national fund for the enrichment and conservation of native forests, which funds will be distributed among the provinces which have finished their Territorial Planning Reports. It is important to say that 30% of this money should be invested in improving technical and public control capacities, and that the other 70% will be devoted to pay to the owners of forests who perform conservation and sustainable management of their resources.

Also, the National Forest Agency is promoting programs and policies for the protection, conservation, and sustainable use of native forests, in coordination with some provincial governments. This is performed through a consultancy mechanism in coordination with provincial governments and several entities connected with the forest area, emphasizing the participation of the communities which live near the natural resource (Minaverry 2013).

In order to focus on the implementation of native forests law, first of all we must analyze the public national budget for each year round, and in the case of Argentina, only between 1 and 2% of the economic incomes were used to finance issues connected with environment protection from 2013 and 2019 (FARN 2019). This applies to all the forest regions of Argentina including Patagonia.

In particular, in 2019, the financial resources devoted for the environmental area represented 0.3% of the whole national budget and this should have been used to cover the National Fund for Conservation of Native Forests regulated in the law. From 2013 until 2019, the budget assigned for the protection and conservation of native forests was always less than what was regulated by law (FARN 2019; Minaverry and Gally 2012).

Also, there is a big difference between the amount of money provided by the National State for the implementation of National Law 25,080 of implanted forests which was fifty times more than the one of native forests during 2016 (FARN 2019). This shows off which is the kind of public policy regarding the protection of native forests and its ecosystem services.

In some provinces of Argentina, there are some challenges in connection with the access to public funds regulated by the National Law of Native Forests 26,331. This situation complements with the existence of a complex bureaucracy system which goes against the implementation of the National Forests Law and its delays regarding payments (Peixotto et al. 2019: 329).

Section 5 of National Law 26,331 defines environmental services as:

"The tangible and intangible benefits, generated by the ecosystems of the native forest, which are necessary for the cooperation and survival of the natural and biological system as a whole, and needed to improve and secure the quality of life of the inhabitants of the nation who benefit from those resources. Among others, the main environmental services that the native forests provide to society are:

- Water regulation.
- Conservation of biodiversity.
- Conservation of the soil and of the quality of water.
- Limit greenhouse gas emissions.
- Contributions to the diversity and landscape beauty.
- Defend cultural identity."

Nonetheless, neither of these laws on environmental or ecosystem services provide policies on criminal penalties, a pending debt in the region due to the existence of high levels of deforestation, and risk of losing natural and cultural biodiversity (Minaverry 2016).

The data gathered denotes the weakness of laws and public policies on ecosystem services in Argentina, due to the fact that there is no political decision about the obligation to pay for these services and any legal parameter pre-established, so the economic value of ecosystem services is arbitrary and based on subjective judgment.

The scarcity of legal rules that refer to the management of ecosystem services hinders their implementation and contrasts with the explosion of regulations on other environmental aspects to the detriment of the effectiveness of the law.

In addition, the deficiencies in the implementation of the ecosystem approach in Argentina are closely related to the structure of Argentine federalism in environmental matters. The Federal State's natural resources jurisdiction is limited because the Constitution establishes that provinces have the exclusive power over their natural resources.

In consideration of the existing cultural diversity in Argentina, we highlight the importance of integrating biodiversity with socio-diversity.

2.3 Public Policies Related with Ecosystem Services Protection and Other Regulation

National and provincial environmental agencies developed different strategies in order to protect native forests and its ecosystem services, such is the case of the Forests Assessment System, Deforestation Alert System, First and Second National Inventory of Native Forests and Forests Statistical National Program.

In this regard, we will describe a selection of relevant public policies related to ecosystem services and native forests protection in Argentina.

The first one is Resolution 267/2019 of the National Environmental Ministry which approved the national plan of restoration of native forests which promotes restoration, recuperation, and rehabilitation of native forests in Argentina.

The main purposes of the plan are the following: Restore the functional processes of native forests and biodiversity finding out the main information about the more affected areas and fixing strategies for floods, soil and forests recuperation, for biological corridors and fire areas, promote the reduction of degradation factors, greenhouse gases, and education about the environment and the importance of native forests and of the different social actors who interact with them.

One of the main issues is that this Resolution states that restoration should be related to socio-economical processes and to promote participation of social actors and the acknowledgment of cultural diversity and multi-ethnicity.

The public agency which will apply this Resolution focuses on: Updating the national plan of restoration of native forests, to develop a monitoring system with indicators, elaborate a nursery garden to produce native plants and encourage the development of financial and market instruments, promote the restoration of native forests in connection with the productive potential of recuperation through the national program of protection of native forests. Each province must prepare its own Plan of Restoration of Native Forests and adhere to the National Resolution.

The second one is the National Plan of Forests and Climate Change of 2017, which is an instrument of public policy which includes the national strategy for the reduction of emissions of deforestation and degradation of Argentina (REDD+). This document is an operational management tool which aims to reduce emissions, to raise the capture of greenhouse gases, and to reduce vulnerability of native forests and of its communities. This plan was elaborated by the National Agency of Climate Change, the academia, NGO's, the private sector, small producers, and indigenous communities.

The financial proposal for Argentina includes the following tools: Action Plan REDD+, National Reference level of forests emissions, National System of Monitory of native forests, and REDD+ safeguards. These are the environmental, social and governance guarantees used to avoid negative impacts and to promote benefits in the territories.

The countries who comply with the previous requirements and reduced its emissions through the implementation of REDD+ can access the payment upon results of the Green Fund for the Climate.

In these policies, it is particularly important that the specific roles of the different governmental levels (national, provincial, and municipal) are clearly defined. The role of the local community in connection with the administration and management of forests must be developed here at this stage. Active participation is a relevant tool to make these actions and policies work out.

Poverty fighting should also be in the South American governments agenda, in the first place to increase people's quality of life and to protect the environment, and valuable forests and ecosystem services.

In the legal arena, the province of Neuquén in Argentina is the only one which regulated the category of "protected indigenous territory" in the framework of protected areas legislation (Section 5, Provincial Law 2342) (Minaverry and Martínez 2018). Also, National Section 4 of the National Law 22,351 defines National Parks stating that in these areas, it is forbidden to perform all economic exploitation except for tourism. Section 6 established that the infrastructure used for tourism could only be essential. The proposal regarding these activities should be presented by the National Administration's Park and include a cause which relies on a general interest and assure that the ecosystems of these areas will be preserved.

One of the purposes of protected areas acknowledged by this regulation is to preserve ecosystem services, but it could be combined with tourism which is considered as an opportunity of enjoyment of aesthetic and cultural values and also the spiritual experience related to the environment. This should also respect cultural integrity and ecological systems which are essential to the communities nearby (Minaverry and Martínez 2018).

In relation with this, National Tourism Law 25,997 regulates it as a socioeconomic, strategic, and essential activity for national development. It also states that this law should incorporate the necessary mechanisms to create, preserve, protect, and use resources and of national tourist attraction in order to achieve a sustainable development (Minaverry and Martínez 2018).

It is important to say that there were previous bills of laws regarding the regulation of ecosystem services at the national level in connection with wetlands, water courses and in some cases with forests. These are the following: (a) environmental services payment supplied by native forests systems in the province of Chaco (Senate file 1417-s-2007), (b) national compensation mechanism of environmental services provided by protected areas managed by the province of Misiones (Senate file 1417-s-2007), and (c) especial regime of environmental services compensation proposed by the representatives of the provinces of Neuquén, La Pampa, Tierra del Fuego, San Luis, and Río Negro (Congress file 1385-d-2011). Only one of the bills of laws relates to Patagonia. However, this region has a high-quality level of biodiversity which should be legally protected.

2.4 Opportunities to Protect Ecosystem Services in Practice

There are several opportunities and troubles regarding the implementation of ecosystem services as a tool, especially in the legal area.

Some authors state that in Argentina, the concept of ecosystem services was launched by international projects which provided funding and in some cases these projects produced changes locally which afterward were translated into public policies and regulations (Peixotto et al. 2019).

In this context, one of the possible ways to incorporate the ecosystem services tool in the governance process is through the acknowledgment and regulation of cultural ecosystem services and to relate them with a sustainable touristic activity.

Additional regulation should be enacted to fill legal gaps and to incorporate the ecosystem approach which is limited in Argentine environmental legislation.

Also, the scarse implementation of environmental courts in some Latin American countries goes against the citizen's freedom and democracy, due to the need of existence of a specialized judicial system. The difficulties arise because there is no ideal model of court to be applied to all the countries of this region, and each case must be analyzed and studied in an independent way.

At present, there are only a few environmental courts (in the province of Jujuy) in Argentina. The Argentine venture is still on its early stage because there are only two Resolutions adopted by Argentina's Supreme Court of Justice, whereby environmental agencies were created and there have been recent efforts toward the creation of specialized courts. Moreover, according to Argentina's case law, the courts that hear these lawsuits, more frequently, are Administrative Courts, Civil and Commercial Courts. To substitute the material absence of specialized courts, the attorneys themselves require technical reports (which in many cases are provided by experts from other fields) (Minaverry 2015). However, when surveying the whole of the case law, we find evidence of the problems and limitations that these aspects and situations create.

Resolution N° 1/2014 enacted by the Supreme Court of Justice on February 11 2014, may have changed the course of the future and probable institutionalization. This regulation supports the creation of an Office for Environmental Justice alleging that "it is vital to have a jurisdiction and independent judicial instances to set in motion the development and the enforcement of environmental law, and the members of the Judicial Power, together with those who contribute to the judicial service at national, regional, and worldwide level, are crucial associates for the promotion, compliance, putting forward and enforcement of the national and international environmental law" (Court Agreement N° 1/2014) (Minaverry 2015).

This office operates inside the National Supreme Court of Justice and has as main functions to follow up legal actions related to environmental law, to connect with similar agencies at a national and international level and to coordinate and manage training programs with other branches of the State and with international environmental organizations related to Justice.

Also the environmental office should take care of the following areas which are directly related to the level of efficiency in the application of environmental law and especially of the national native forests law, to training and to compilation of data and research (in collaboration with universities and research and educational centers).

In the second place, Resolution 8/2015 enacted by the Supreme Court of Justice created the Environmental Trials Secretary for the processing of files related to environmental issues. They incorporated specialized jurists to collaborate, provide doctrine and to analyze case law. This can be understood as a previous stage to the environmental justice implementation. The main reason for this was that the State must promote the instrumentalization of tools which help to achieve community's needs, to access efficiently to environmental justice and to coordinate massive collective trials management.

In the province of Neuquén (Northern Patagonia), there is a bill of laws which was initiated in 2018 (12,312, D-931/18) which regulates the creation of an environmental jurisdiction.

However, the static role of judges all along the legal process is usually criticized because they usually do not move to the place where the facts occur and have a passive attitude which in some cases might change legal resolutions.

Environmental rising tensions are currently fixing the need to provide administrative and judicial procedures, to stop damaging effects of human behavior, and of the risks created by them. Notwithstanding, at present, there are only a few cases of environmental courts functioning in Latin America.

National and provincial legislation have incorporated the concept indirectly, and this has not contributed to promoting the creation of public policies, nor did it foster its treatment in the political agenda (since there is no relevant legislative framework that regulates the subject).

2.5   Final Remarks

There are several setbacks which relate to environmental law ineffectiveness regarding the implementation of valid regulation of native forests protection and of ecosystems services.

At present, the regulatory and jurisprudential development related to ecosystem services is emerging and has no relevance. We expect a prompt incorporation and elaboration of public policies related to these issues. Nevertheless, the payment mechanism of native forests law can be considered as one important exercise in Argentina which is connected to ecosystem services compensation (Peixotto et al. 2019).

It is especially important that the valid and existing laws are rigorously applied by the corresponding public authorities (administrative and judicial), because if not its mere existence is useless. The superposition of regulations is common in Argentina, but it can be solved through the amendment of valid legislation, but mainly to avoid the lack of access to environmental rights.

It should be considered that in most cases, there is no need to regulate more the use of natural resources or the protection of forests or ecosystem services. In practice, there are key issues such as management and governance which are most relevant and should be developed further.

We have noticed a lack of public policies related to the present issue jointly (Environmental Law and Social Anthropology), and it will be necessary that if adopted in the future, they emerge from the contributions made by expert researchers, from the public and private sector, and the community, by combining their scientific knowledge, their social institutional and financial expertise, in order to be able to face the complexity of the socio-ecological systems that we have analyzed (Balvanera et al. 2012).

We can highlight a new and fundamental socio-environmental principle entitled "*in dubio pro natura*" which was incorporated in 2019 by the Supreme Court of

Fig. 1 Forests in Ushuaia (I), Tierra del Fuego, Argentina. *Source* Minaverry (2019)

Fig. 2 Forests in Ushuaia (II), Tierra del Fuego, Argentina. *Source* Minaverry (2019)

Justice of Argentina and set a landmark. The judicial case "Majul Julio Jesús against the Municipality of Pueblo General Belgrano and others about environmental protection" focused on the existence of a negative alteration of the environment connected with wetlands and hydric basins.

Finally, and especially regarding Covid-19, research concerning health and environmental areas should be valued, sponsored, and incorporated in the political agendas, not only taking into consideration technical approaches but also the social area connected with the communities related with the forests (Figs. 1 and 2).

References

Aguiar S et al (2018) ¿Cuál es la situación de la Ley de Bosques en la Región Chaqueña a diez años de su sanción? Revisar su pasado para discutir su futuro. Ecol Austral 28:400–417

Balvanera P et al (2012) Ecosystem services research in Latin America: the state of art ecosystem services. Ecosyst Serv 2:56–70

Castelnuovo N (2019) Representaciones e ideologías de ONG confesionales en el Chaco Argentino. Revista Antropologías Del Sur Año 6(11):39–61

Castilla MI (2017) De lo global a lo local: políticas públicas y surgimiento de organizaciones étnicas en Pampa del Indio, Chaco. Papeles De Trabajo 34:29–44

Castilla MI (2018) Territorios y fronteras: procesos de apropiación del espacio simbólico y geográfico en las comunidades indígenas de Pampa del Indio, Chaco, Boletim do Museu Paraense Emílio Goeldi. Ciências Humanas 13(3):541–560

Castilla M, Weiss ML, Engelman JM (2019) Transformaciones socioeconómicas, migración y organización etnopolítica rural-urbana entre la Región Chaqueña y la Región Metropolitana de Buenos Aires. Cuadernos de Antropología Social (49)

Costanza R et al (1997) The value of the world's ecosystem services and natural capital. Nature 387:253–260

Costanza R, de Groot R, Braat L, Kubiszewski I, Fioramonti L, Sutton P, Farber S, Grasso M (2017) Twenty years of ecosystem services: how far have we come and how far do we still need to go? Ecosyst Serv 28:1–16

Daily G (1997) Nature's services: societal dependence on natural ecosystems. Island Press, Kindle version

Engelman JM (2019) Indígenas en la ciudad: articulación, estrategias y organización etnopolítica en la Región Metropolitana de Buenos Aires, Argentina. Quid 16. Revista Del Área De Estudios Urbanos 11:86–108

Ferro M, Minaverry C (2019) Aportes normativos, institucionales y sociales a la gestión del agua y el enfoque ecosistémico en la Cuenca del Río Luján, Argentina. Revista de Derecho 20:25–55. Facultad de Derecho, Universidad Católica del Uruguay. Montevideo

Flores Klarik M (2019) Agronegocios, pueblos indígenas y procesos migratorios rururbanos en la provincia de Salta, Argentina. Revista Colombiana De Antropología 2(55):65–92

Fundación Ambiente y Recursos Naturales (FARN) (2019) El presupuesto ambiental entre 2013 y 2019: una historia de desfinanciamiento. https://farn.org.ar/wp-content/uploads/2020/03/FARN_El-presupuesto-interanual-entre-2013-y-2019-1.pdf. Last access 9 Oct 2020

Infobae—Sociedad—Salta: una nena wichi de un año y 9 meses murió por desnutrición 16 de Diciembre de 2020. https://www.infobae.com/sociedad/2020/12/16/salta-una-nena-wichi-de-un-ano-y-9-meses-murio-por-desnutricion/. Last access 20 Dec 2020

Maidana CA (2013) Migración indígena, procesos de territorialización y análisis de redes sociales. REMHU, Rev Interdisc Mobil Hum 41(21):277–293

Mancinelli G (2019) Territorialidad y Educación superior en comunidades wichí del noreste salteño. Unpublished doctoral thesis, Facultad de Filosofía y Letras (University of Buenos Aires)

Millennium Assessment/MA (2005) Ecosystem and human well-being: synthesis. Island Press, Washington D.C

Minaverry CM (2013) The new law for the protection of native forests in Argentina. Revista Ars Boni Et Aequi, Año 9(1):237–246

Minaverry CM (2015) El avance de la implementación de los tribunales ambientales en América Latina. Revista Gestión y Ambiente 2(18):95–108

Minaverry CM (2016) El Derecho Penal Ambiental a la luz de un reciente precedente jurisprudencial sobre desmontes ilegales en la Provincia de Salta, Argentina. Errepar, Buenos Aires

Minaverry CM (2020) El reconocimiento de los servicios ecosistémicos culturales en el ámbito científico del Derecho Ambiental y de las Ciencias Sociales. Aplicación de sus contribuciones al caso de la protección jurídica de los bosques nativos en Norpatagonia argentina y chilena. Revista Lex de la Facultad de Derecho y Ciencias Políticas 18(25):109–137

Minaverry C, Gally T (2012) La implementación de la protección legal de los bosques nativos en la Argentina. Revista Pensamiento Jurídico 35:253–278

Minaverry C, Martínez A (2018) Los servicios ecosistémicos culturales y su interrelación con el Derecho Ambiental. El caso de Argentina. 56° Congreso Internacional de Americanistas, in: M. Alcántara, M. García Montero and F. Sánchez López (Coordinators), Estudios sociales. Memoria del 56° Congreso Internacional de Americanistas. Editorial Universidad de Salamanca, Salamanca, pp 534–545. Spain

Murphy A, Grosso S (2012) Impactos socio-territoriales del avance de un frente agropecuario: Estudio realizado en el Este de la provincia de Santiago del Estero. In: VII Jornadas de Sociología de la Universidad Nacional de La Plata, 5 al 7 de diciembre de 2012, La Plata, Argentina. Argentina en el escenario latinoamericano actual: Debates desde las Ciencias Sociales. La Plata: Universidad Nacional de La Plata. Facultad de Humanidades y Ciencias de la Educación. Departamento de Sociología

"Proyecto de monitoreo de deforestación en el Chaco Seco". http://monitoreodesmonte.com.ar/. Last access 9 Oct 2020

Naharro N, Álvarez AL (2011) Estudio de caso: acaparamiento de tierras y producción de soja en Territorio Wichi, Salta—Argentina (Berlín: Brot für die Welt/Buenos Aires: Asociana-Argentina). http://redaf.org.ar/wp-content/uploads/2011/12/Estudio-de-Caso-Produccion-de-soja-en-territorio-Wichi-Salta. Last access 9 Oct 2020

National and Environmental Sustainable Ministry (2018) Environmental status report. Last access 9 Apr 2021

Paruelo JM, Guerschman JP, Piñeiro G, Jobbágy EG, Verón SR, Baldi G, Baeza S (2006) Cambios en el uso de la tierra en Argentina y Uruguay: marcos conceptuales para su análisis. Agrociencia 2(X):47–61

Paz ML (2015) Crisis de los sistemas productivos agrarios tradicionales. El acceso a los recursos y la dinámica de la población rural en tiempos de 'bovinización'. RUNA, Archivo para las Ciencias del Hombre 36(1):109–124. Cruz del Eje, Córdoba, Argentina.

Paz ML, Fleitas K (2019) Del líquido vital: entre la escasez y el valor de uso en la reproducción social campesina. Mundo Agrario 20(44). Cruz del Eje, Córdoba, Argentina.

Peixotto J, Godfrid J, Stevenson H (2019) La difusión del concepto de servicios ecosistémicos en la Argentina: alcances y resistencias. Revista SAAP: Sociedad Argentina De Análisis Político 13(2):313–340

Peretti P (2020) Pandemia y Latifundio, Página 12. Suplemento y Sección "Economía", 27 de mayo de 2020. https://www.pagina12.com.ar/268385-pandemia-y-latifundio. Last access 9 Oct 2020

Quétier F, Tapella E, Conti G, Cáceres D, Díaz S (2007) Servicios ecosistémicos y actores sociales Aspectos Conceptuales y Metodológicos Para Un Estudio Interdisciplinario. Gaceta Ecológica 84–85:17–26

Scheinkerman de Obschatko E, Foti M del P, Román M (2007) Los pequeños productores en la República Argentina. Importancia en la producción agropecuaria y en el empleo en base al Censo Nacional Agropecuario 2002. Secretaría Agricultura, Ganadería, Pesca y Alimentos. Buenos Aires

Schmidt M (2017) Crónicas de un (Des)Ordenamiento Territorial. Disputas por el territorio, modelos de desarrollo y mercantilización de la naturaleza en el este salteño. Buenos Aires: Teseo.   http://antropologia.institutos.filo.uba.ar/sites/antropologia.institutos.filo.uba.ar/files/info_covid_2daEtapa.pdf. Last access 10 Oct 2020

Sebastián, Valverde C M, Minaverry G, Stecher (2021) Examining the "Forest Law" in Los Lagos Argentina Through the Lens of Mapuche Organisations. Journal of Intercultural Studies 42(2) 160-176 10.1080/07256868.2021.1883570

Tamagno L, Maidana C (2011) Grandes urbes y nuevas visibilidades de la diversidad. Revista Brasilera De Estudos Urbanos e Regionais 1(13):51–61

Environmental Policies and Territorial Conflicts in Argentina—From the Deforestation of Native Forests to Agrochemical Spraying in the Province of Salta

Mariana Schmidt

Abstract Since the end of the twentieth century in Argentina, a new model of agricultural development has been established, with increasing environmental, social, and sanitary impacts. Based on the information collected and constructed during our doctoral research and the survey of new sources (documentary, statistics, cartographic, hemerographic, legislative, and bibliographic), this chapter tries to unravel the advances and challenges towards a fairer environmental, social and territorial scenario in the context of the implementation of National Law of "National Forests," enacted in 2007. We focus in the case of the province of Salta, in Northern Argentina: a geographical space involved in processes of expansion of extractive frontiers over territories that emerge as living spaces for indigenous and rural populations, which have been historically marked by processes of sub-alternation and situations of deep environmental and social injustice.

Keywords Public policies · Environmental · Forests · Regulations

1 Introduction

The appropriation and distribution conflicts that exist at the core of the Latin American societal matrix must be analyzed in their economic, ecological, and cultural dimensions (Alimonda 2011; Escobar 2011; Martínez Alier 2006). These conflicts cross all realms of human and non-human existence, and they are expressed as social and environmental injustices, and in the center of the debates lies the defense and claims over "the common" (Laval and Dardot 2015).

It is not just a matter of uneven distribution of existing risks and resources. The long-standing processes that caused and continue to perpetuate unequal distribution

M. Schmidt (✉)
Consejo Nacional de Investigaciones Científicas y Técnicas (CONICET), Facultad de Ciencias Sociales, Universidad de Buenos Aires, Buenos Aires, Argentina

© The Author(s), under exclusive license to Springer Nature Switzerland AG 2021
C. M. Minaverry and S. Valverde (eds.), *Ecosystem and Cultural Services*, The Latin American Studies Book Series, https://doi.org/10.1007/978-3-030-78378-5_2

and appropriation must be considered. Other relevant dimensions must also be examined: the lack of recognition of affected communities, the procedural inequalities that marginalize these communities in the participatory instances of design, implementation, management, and assessment of public and/or private policies and interventions that concern them, and the restriction of functioning capacities (individual, but mainly collective) in these contexts (Acselrad et al. 2009; Blanchon et al. 2009; Boelens et al. 2011; Carruthers 2008; Harvey 2018; Schlosberg 2011; Schlosberg and Carruthers 2010).

The map of environmental conflicts in Argentina has been expanding in all directions since the end of the twentieth century (Merlinsky 2013, 2016, 2020; Svampa and Viale 2014) and is particularly relevant in the light of changes linked to the consolidation of agribusiness (Giarracca and Teubal 2008; Gras and Hernández 2013). This model combines the export-oriented monoculture of genetically modified crops with intensive agrochemical use. It entails a land agriculturalization process, along with the "soyzation" of crops. At the end of the 1970s in Argentina, less than one million hectares were allocated to the production of soybean, whereas by the 2012/13 crop year, the cultivated area was 20 million hectares (more than half of the country's agricultural land).[1] The numbers showing the increase of cultivated land correspond to an even higher expansion of the use of agrochemicals in the productive process.[2]

As a result, conflicts over the loss and deterioration of common goods, over the eviction and destruction of peasant and indigenous ways of living, and the health and environmental impacts due to the chronic exposure to spraying have gained public visibility. This chapter draws from information collected and constructed during my doctoral research (Schmidt 2017) and considers new sources (documents, statistics, maps, newspapers, legislation, and bibliography). It focuses on the province of Salta, in the Argentine north: a geographic area involved in processes of expansion of extractive frontiers with high environmental and sanitary impact upon the populations that live in the territory and use its resources.

2 The Case of the Province of Salta, Argentina

The expansion of the agricultural frontier from the traditional *Pampa* region toward the North of the country was driven by developments in biotechnology, a dynamic real estate market and a favorable international context. In the early 1990s in the province of Salta, 100,000 hectares (ha) were sowed with soybean. This number increased to 600,000 ha in the 2010/2011 crop year (439,512 ha in 2018/19). Other crops also increased in the province, such as maize (287,333 ha in 2018/19) and the

---

[1] http://www.agroindustria.gob.ar/datosabiertos/.

[2] The use of agrochemicals has increased since the 1990s (about 30 million liters/kg per year), up to over 500 million liters/kg in the latest crop years. https://naturalezadederechos.org/525.pdf.

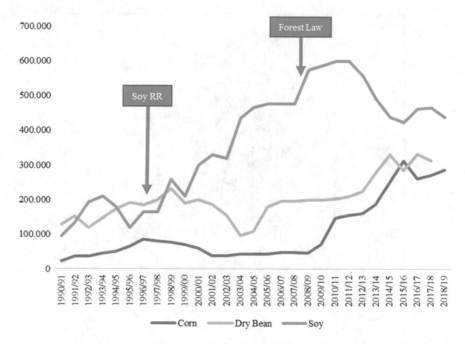

Fig. 1 Soybean, corn and dry poroto extension in the province of Salta, Argentina. Campaigns 1990/91–2018/19. *Source* Author's elaboration based on http://www.agroindustria.gob.ar/datosabiertos/

dry bean (313,483 ha in 2018/19). In addition to crops (that currently cover over 1,200,000 ha), corporate cattle farming has driven land-use changes in more arid regions (stock increased from over 500,000 heads of cattle at the beginning of 2000 to nearly 1,500,000 in 2019) (Fig. 1).[3]

To a large extent, the increase of agricultural and livestock farming areas was made at the expense of intense deforestation and biodiversity loss. In the Dry Chaco region that comprises parts of Argentina, Bolivia, and Paraguay, 15,800,000 ha of natural habitats were transformed between 1976 and 2012 (Vallejos et al. 2015). On the national level, 11,875,852 ha were deforested between 1976 and 2018. Salta is one of the jurisdictions with the highest proportions of loss of native forests (17.3%) (Fig. 2).[4]

In this context, in 2007, the National Law 26,331 on Minimum Standards for the Environmental Protection of Native Forests was debated and enacted. The law mandated that the province conduct its Territorial Planning of Native Forests (OTBN, for its initials in Spanish) with the sanction of a rule and through a

---

[3]The number of cattle increased from over 500,000 heads at the beginning of 2000 to approximately 1,400,000 in 2019, and the province has two of the country's largest feedlots.
[4]http://monitoreodesmonte.com.ar/.

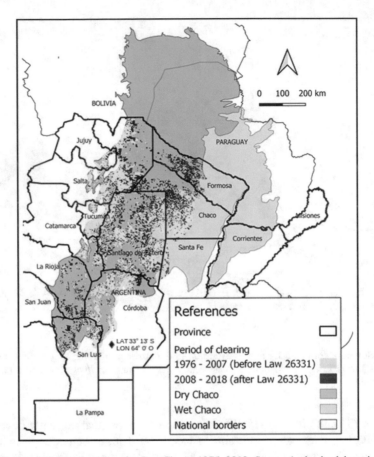

Fig. 2 Deforestation in the Argentine Dry Chaco. 1976–2018. *Source* Author's elaboration based on http://monitoreodesmonte.com.ar/

participatory process, within one year at the most. OTBNs classified forests into three categories—high (red), medium (yellow), and low (green) conservation value —according to ten criteria of environmental sustainability. The passing of the "Forest Law" was the result of a lengthy public debate about the protection of remaining forest mass in Argentina. However, it was also the starting point for the implementation of an environmental policy that, at its different levels and scope of application, has rekindled the tension—usually presented as an irreconcilable dichotomy—between nature conservation (backwardness) and production (progress) (Schmidt 2017). During the process, contradictions have surfaced: when the discussion passed from the national scale to the provincial jurisdictions—the on-site application authorities of environmental regulation—discourses and actors were reconfigured, frequently activating new conflicts and controversies (Langbehn 2015) (Fig. 3).

Fig. 3 Deforestation to use lands for agriculture in the district of General San Martín, Salta, Argentina. *Source* Photograph taken by the author of this chapter, 2007

The case of Salta is particularly significant: Not only it is one of the jurisdictions with the largest area of native forests in Argentina but also is one of the areas with the highest potential for agriculture and livestock farming. Besides, the province has a large volume of contested land in hands of indigenous and peasant dwellers, and it was a precursor both in deforestation rates and in the speedy implementation of the OTBN plan. A participatory process for the definition and zoning of forest protection areas took place at the beginning of 2008. Many workshops, zonal, and sectorial meetings took place with the participation of a broad and diverse set of actors: government and non-government officials and technical experts, members of forestry and farming producers associations, indigenous organizations and movements, and *criollo* families. However, this active participation did not translate into a law responding to the different sectors' demands: the legislature that was responsive to the interests and pressure of large farming producer associations, modified several of the original project's articles, reducing the scope of forest protection and weakening the safeguards that it gave indigenous peoples (Fig. 4).

Days after the enactment of OTBN Provincial Law 7543, a group of indigenous and peasant organizations presented a protective action to the Federal Supreme Court of Justice. They obtained an injunction, which paralyzed all deforestation activities in four departments of the province's Northwest. In 2009, the Salta government eventually sanctioned the Regulatory Decree 2785, with an OTBN map that zoned 5,400,000 ha of forest under yellow category and 1,300,000 ha under red category (65% and 16% of total forest area, respectively), whereas 19% were defined as areas for productive use (Fig. 5). However, in the following years, the

Fig. 4 Transportation of wood in Embarcación, district of General San Martín, province of Salta, Argentina. *Source* Photograph taken by the author of this chapter, 2008

sanctioning of the norm, the province continued to lose forest cover in authorized and prohibited areas (Aguiar et al. 2018; Schmidt 2017; Langbehn 2015; Seghezzo et al. 2011).

The different OTBN proposals revealed various actors' projects and intentions of dominating and/or appropriating the territory in Salta. These proposals were, ultimately, contested development models and form of nature valuation (Schmidt 2017). Thus, the process of OTBN design, formulation, and implementation speaks of the historic conflicts inscribed in these territories. In addition, it reveals the existence of actors with a higher capacity to impose their territorial project and appropriate the positive balances of environmental and territorial expropriation and dispossession.

The expansion of the extractive frontiers must be understood in relation to the historic conflicts over land use, tenure, and ownership (Castelnuovo 2019; Slutzky 2005). In the Chaco region of Salta, about 3,000,000 ha are inhabited and/or claimed by indigenous communities (2,674,398 ha) and small producers (1,200,964 ha, of which 976,265 ha overlap with indigenous lands) (Salas Barboza et al. 2020). Large transactions of lands have taken place in over 1,600,000 ha in the 2000–2017 period, and almost 50% are involved in conflicts regarding deforestation, environmental damage, and/or tenure (Agüero et al. 2019). In addition, land grabbing

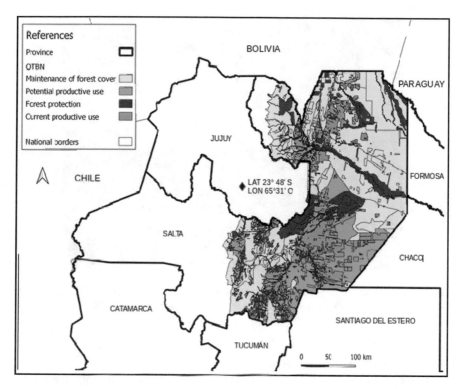

Fig. 5 OTBN zoning in the province of Salta. *Source* Author's elaboration based on http://www.idesa.gob.ar/

Fig. 6 Grains selected in silos in an agro-industrial company in the district of General San Martín, province of Salta, Argentina. *Source* Photograph taken by the author of this chapter, 2019

Fig. 7 Graffiti against deforestation in Embarcación, district of General San Martín, province of Salta, Argentina. *Source* Photograph taken by the author of this chapter, 2008

has functioned as a mechanism to expropriate and export the region's available freshwater (Agüero et al. 2016) (Fig. 6).

Overall, the processes of land grabbing and appropriation of the commons in the Chaco region of Salta do not take place on the backdrop of "empty" scenarios, as some hegemonic narratives have contended. Instead, they are superimposed upon and/or re-signify long-standing injustices and inequalities. The different actors involved use, appropriate, and distribute the available resources in profoundly different ways, now and in the past. The indigenous, peasant, and rural population that lives there consider forests and rivers as common goods that provide their habitat, refuge, and food, where they produce and reproduce daily.

Also, not only is there an unfair distribution in the access and use of water, forests, and land but also high levels of degradation and pollution must be considered. Indeed, the lands that indigenous and peasant populations inhabit are degraded, insufficient in size, and increasingly cornered between large, cultivated areas. The forest (locally named as "*monte*"), a primary source for daily gathering, hunting, artisanal production, and the provision of firewood, is becoming gradually scarcer, making it therefore necessary to cover increasingly longer distances. In addition to the lack of basic water and sanitation services and the restricted access to safe water in a region marked by scarcity, the sources of superficial and underground water available for human and/or domestic consumption are also contaminated (Fig. 7).

The preliminary results of a survey about conflicts related to the use, application and/or storage of agrochemicals in the province of Salta (a total of 166 cases surveyed up to March 2020), highlight significant points for the 1999–2020 period

(Schmidt et al. 2019). The departmental distribution reveals that 82% of cases are located in four of the jurisdictions most affected by the expansion of large-scale agribusiness production (in order of importance: San Martín, Anta, Orán, and Metán). The mainly affected population is from the urban periphery (35.4%) and indigenous people (36.1%). Regarding the defendants, they are mostly private actors (large businesses, producers, and applicators); however, municipal or provincial authorities are also held responsible for control and sanction activities. The main detected situations are linked to "bad" agriculture practices (pesticide spraying in fields adjacent to homes and/or schools, machinery, and container storage containers in urban areas, etc.), environmental (flora and fauna mortality, water, air, and soil pollution), and sanitary impact (intoxications, congenital anomalies, diseases, etc.).

These conflicts, although latent or in the process of emerging, are to a large extent invisible and/or subsumed under other more profound historical injustices (such as conflicts over land). Public demonstrations and claims are still infrequent and the majority hardly transcends the local sphere.[5] Discussions and decisions still occur within the framework of the current agricultural development model, excluding the participation of the affected populations, whose knowledge and experiences are undervalued in relation to technical and expert knowledge. The situation is further exacerbated by the historic material and symbolic dispossession of peasant and indigenous communities, as well as that of the small towns that have been cornered by large-scale agricultural ventures and that, on a day to day basis, suffer the impacts of agrochemicals that affect their bodies and their life spaces.

However, this does not mean that the environmental policy of forest protection has not had effects. Indeed, a series of "productivities" can be acknowledged at the juridical, political-institutional, territorial, and social levels (Melé 2016; Merlinsky 2013): the creation of new regulations along with the appropriation and/or mobilization of existing laws; the unfolding of processes of juridical qualification of space as a consequence of the territorial zonings foreseen by OTBN laws; the higher visibility, on the local and national spheres, of conflicts (environmental and territorial), of the actors involved and their opposing interests and valuation languages; the controversies about the environmental and/or sanitary risks resulting from deforestation and spraying, along with the unfolding of arguments that contest the truths of hegemonic science; the mobilization and/or resignification of new meanings associated to the contested territories (environmental, but also cultural and related to identity); the emergence of actors and the generation of strategic alliances on the different scales of analysis; the (relative) strengthening of environmental organisms at the national and provincial levels, of their human and material resources; and the tensions at the interjurisdictional levels and over the control of territories, to mention some of them.

---

[5]It is noteworthy to mention the case of Antillas (a town in the department of Rosario de la Frontera), the only case in the province of Salta where after presenting a legal protective action, a group of residents obtained an injunction from the Judicial Power prohibiting aerial spraying at less than 1500 m from homes, and 300 m in the case of ground application.

3 Final Remarks

The federal forest protection policy was gradually expressed in provincial laws through their OTBN maps, institutional arrangements, and the regulations enacted over the years. However, this did not mean the end of the disputes over the "commons": Profound injustices and development matrixes remain unproblematized and unquestioned (Schmidt 2019). As Schlosberg and Carruthers (2010) highlighted, the indigenous peoples, in their struggle against environmental degradation and for territorial rights, embrace more pluralistic notions of environmental justice. In doing so, they incorporate broader claims about the preservation of practices, cultures, knowledge, and languages, the continuity of their ways of relating to nature, and the protection of household economies. It is necessary to permanently commit to other productive (and life) alternatives, socially and environmentally more just, through an engaged and intercultural "dialogue between knowledges" (De Sousa Santos 2009) with the local populations and their experiences and interests.

References

Acselrad H, Mello C, Bezerra G (2009) O que é justiça ambiental? Garamond, Rio de Janeiro

Agüero JL, Salas Barboza A, Venencia C, Müller M, Seghezzo L (2016) Grandes transacciones de tierras como mecanismo de apropiación y exportación de agua en la región del Chaco salteño. ASADES 20:37–48

Agüero JL, Venencia C, Talamo A, Salas Barboza A, Díaz Paz W, Sajama J (2019) El fenómeno de las grandes transacciones de tierras en la región del Chaco de la provincia de Salta, Argentina. In: Simón M et al (eds) Grandes transacciones de tierra en América Latina: sus efectos sociales y ambientales, pp 22–36. Fundapaz, Buenos Aires

Aguiar S, Mastrángelo M, García Collazo MA, Camba Sans G, Mosso C, Ciuffoli L (2018) ¿Cuál es la situación de la Ley de Bosques en la Región Chaqueña a diez años de su sanción? Revisando su pasado para discutir su futuro. Ecología Austral 28:400–417. https://doi.org/10.25260/EA.18.28.2.0.677

Alimonda H (2011) La colonialidad de la naturaleza. Una aproximación a la Ecología Política Latinoamericana. In: Alimonda H (Coordinator) La Naturaleza colonizada. Ecología política y minería en América Latina, pp 21–58. CLACSO, Buenos Aires

Blanchon D, Moreau S, Veyret Y (2009) Comprendre et construire la justice environnementale. Annales De Géographie 665–666:35–60

Boelens R, Cremers L, Zwarteveen M (2011) Justicia Hídrica: acumulación de agua, conflictos y acción de la sociedad civil. In: Justicia hídrica: acumulación, conflicto y acción social, pp 13–22. IEP-Fondo Editorial PUCP-Justicia Hídrica, Lima

Carruthers D (2008) Introduction. Popular environmentalism and social justice in Latin America. In: Carruthers D (ed) Environmental justice in Latin America: problems, promise, and practice, pp 1–22. The MIT Press, Londres

Castelnuovo N (2019) Pueblos indígenas y grandes transacciones de tierra en el noroeste argentino. In: Simón M et al (eds) Grandes transacciones de tierra en América Latina: sus efectos sociales y ambientales, pp 53–87. Fundapaz, Buenos Aires

De Sousa Santos B (2009) Una epistemología del sur: la reinvención del conocimiento y la emancipación social. Siglo XXI, Ciudad de México

Escobar A (2011)\ Ecología Política de la globalidad y la diferencia. In: H. Alimonda (Coordinator) La Naturaleza colonizada. Ecología política y minería en América Latina, pp 61–92. CLACSO, Buenos Aires

Giarracca N, Teubal M (2008) Del desarrollo agroindustrial a la expansión del "agronegocio": el caso argentino. In: Mançano Fernandes B (Org) Campesinato e agronegócio na América Latina: a questao agraria atual, pp 139–164. CLACSO, Sao Paulo

Gras C, Hernández V (Coordinators) (2013) El agro como negocio. Producción, sociedad y territorios en la globalización. Biblos, Buenos Aires

Harvey D (2018) Justicia, Naturaleza y la geografía de la diferencia. Traficantes de Sueños, Quito

Langbehn L (2015) Arenas de conflicto y construcción de problemas públicos ambientales. Un análisis de la productividad del caso de la Ley de Bosques y del Ordenamiento Territorial de Bosques Nativos de Salta (2004–2009). Unpublished doctoral thesis, Facultad de Ciencias Sociales, Universidad de Buenos Aires

Laval C, Dardot P (2015) Común. Ensayo sobre la revolución en el siglo XXI. Gedisa, Barcelona

Martínez Alier J (2006) El ecologismo de los pobres. Conflictos ambientales y lenguajes de valoración. Icaria, Barcelona

Melé P (2016) ¿Qué producen los conflictos urbanos? In: Carrión F, Erazo J (Coordinators) El derecho a la ciudad en América Latina. Visiones desde la política, pp 127–157. PUEC-UNAM, International Development Research Center, IDRC/CRDI, México

Merlinsky G (Comp) (2013) Cartografías del conflicto ambiental en Argentina. Ciccus, Buenos Aires

Merlinsky G (Comp) (2016) Cartografías del conflicto ambiental en Argentina II. Ciccus, Buenos Aires

Merlinsky G (Comp) (2020) Cartografías del conflicto ambiental en Argentina III. Ciccus, Buenos Aires

Salas Barboza A, Cardón Pocoví J, Venencia C, Huaranca L, Agüero JL, Iribarnegaray MA (2020) Ten years of contested enforcement of the Forest Law in Salta, Argentina. The role of land-change science and political ecology. J Land Use Sci 15:221–234. https://doi.org/10.1080/1747423X.2019.1646333

Schlosberg D (2011) Justicia ambiental y climática: de la equidad al funcionamiento comunitario. Ecología Política 41:25–35

Schlosberg D, Carruthers D (2010) Indigenous struggles, environmental justice, and community capabilities. Glob Environ Politics 10(4):12–35

Schmidt M (2017) Crónicas de un (Des)Ordenamiento Territorial. Disputas por el territorio, modelos de desarrollo y mercantilización de la naturaleza en el este salteño. Editorial Teseo, Buenos Aires

Schmidt M (2019) (In)justicias ambientales, territoriales y socio-sanitarias en el Chaco salteño, Argentina. Folia histórica del Nordeste (35):7–26. http://dx.doi.org/https://doi.org/10.30972/fhn.0353575

Schmidt M, Grinberg E, Langbehn L, Álvarez A, Pereyra H, Toledo López V (2019) Riesgos e impactos socio-sanitarios de las fumigaciones con agroquímicos en las provincias de Salta, Santiago del Estero y Santa Fe. Final report presented to the Convocatoria a Becas de Investigación "Salud investiga, Dr. Abraham Sonis" 2018, Secretaría de Salud de la Nación.

Seghezzo L, Volante J, Paruelo J, Somma D, Buliubasich C, Rodríguez H (2011) Native forests and agriculture in Salta (Argentina): conflicting visions of development. J Environ Dev 20 (10):1–27. https://doi.org/10.1177/1070496511416915

Slutzky D (2005) Los conflictos por la tierra en un área de expansión agropecuaria del NOA. La situación de los pequeños productores y los pueblos originarios. Revista Interdisciplinaria De Estudios Agrarios 23:59–100

Svampa M, Viale E (2014) Maldesarrollo. La Argentina del extractivismo y el despojo. Katz, Buenos Aires

Vallejos M, Volante J, Mosciaro JM, Vale L, Bustamante ML, Paruelo J (2015) Transformation dynamics of the natural cover in the Dry Chaco ecoregion: a plot level geodatabase from 1976 to 2012. J Arid Environ 123:3–11. https://doi.org/10.1016/j.jaridenv.2014.11.009

Regulatory Progress and Social Perception of the Ecosystem Services Provided by Urban Forests in the Municipality of Luján, Province of Buenos Aires, Argentina

Analía Scarselletta and Elena Beatriz Craig

Abstract Urban centers are complex, intense and dynamic anthropic systems, where unplanned expansion has social, economic and environmental negative consequences. Luján is one of the 135 municipalities that are part of the Province of Buenos Aires, in Argentina. It presents a welfare index of 6.76, positioning itself in the 141° ranking of 528 Argentine cities. A significant amount of work has been found that recognizes, evaluates and describes the potential of the services provided by urban forests to achieve "inclusive, safe, resilient and sustainable cities" as described in the Sustainable Development Goal No. 11 of the 2030 Agenda. However, in Argentina, environmental services have not been regulated autonomously. The purpose of this work is to analyze local regulatory framework for the recognition, assessment and/or protection of the services provided by urban forests; and to explore the social perception about these services. This is a qualitative research with an exploratory, documentary and descriptive aim. Understanding how different actors perceive ecosystem services can be a tool to update local regulation and support decision-making to improve the quality of life of current inhabitants and future generations.

Keywords Ecosystem services · Urban forests · Environmental law · International instruments

A. Scarselletta (✉)
Departamento de Ciencias Sociales and Departamento de Tecnología, Universidad Nacional de Luján, Luján, Argentina

E. B. Craig
Departamento de Tecnología, Universidad Nacional de Luján, Luján, Argentina

© The Author(s), under exclusive license to Springer Nature Switzerland AG 2021
C. M. Minaverry and S. Valverde (eds.), *Ecosystem and Cultural Services*, The Latin American Studies Book Series, https://doi.org/10.1007/978-3-030-78378-5_3

1   Introduction

The New Urban Agenda[1] acknowledges that "cities and human settlements face unprecedented threats from unsustainable consumption and production patterns, loss of biodiversity, pressure on ecosystems, pollution, natural and human-made disasters and climate change and its related risks, undermining the efforts to end poverty in all its forms and dimensions, and to achieve sustainable development." The growth of unplanned cities has a direct impact on sustainability and resilience, beyond the borders of urban areas. Consequently and consistently, regarding the topic we address, in Section 67, the signatory countries among them, Argentina, commit themselves to "promoting the creation and maintenance of well-connected and well-distributed networks of open, multipurpose, safe, inclusive, accessible, green and quality public spaces, to improve the resilience of cities to disasters and climate change, including floods, drought risks and heat waves, to improve food security and nutrition, physical and mental health and household and ambient air quality, to reduce noise and promote attractive and human settlements and urban landscapes and to prioritizing the conservation of endemic species" (United Nations, 2016).

The Intergovernmental Science-Policy Platform on Biodiversity and Ecosystem Services (IPBES) emphasize that the ways to understand, study, value and manage ecosystems vary, depending on how each society views the world and values its ecosystems. Knowing the ways in which nature is valued, we expand our understanding of the benefits it provides. Describing and quantifying nature's contributions to human's quality of life become crucial to motivate, guide and justify the development of policies and management actions (IPBES 2018).

Argentina has adapted the Sustainable Development Goals (SDG)[2] to its national reality, through the work conducted by the National Council for the Coordination of Social Policies. Thus, the National State coordinates the 2030 Agenda and the execution of eight Strategic Goals (with their targets and indicators) set forth in the National Government Plan (Secretariat of Land Planning and Public Work Coordination 2018).

---

[1]On October 20th, 2016, the United Nations Conference on Housing and Sustainable Urban Development (Habitat III) held in Quito, Ecuador approved the New Urban Agenda. The United Nations General Assembly endorsed the New Urban Agenda on its sixty-fourth plenary session of its seventy-first session, on December 23rd, 2016.

[2]In the United Nations Conference on Sustainable Development, Rio + 20 Summit, held in 2012, the 2030 Agenda was drafted, and 17 Sustainable Development Goals (SDG) were established. These goals represent a universal call for the adoption of measures aimed at working towards sustainable development through a holistic perspective, focusing on the economic, social and environmental dimensions (United Nations 2018).

Urban forests[3] directly or indirectly contribute to complying with several SDGs, particularly SDG 11—"Make cities and human settlements inclusive, safe, resilient and sustainable"—and SDG 15—"Promote the sustainable use of terrestrial ecosystems, control desertification, halt and reverse land degradation and halt biodiversity loss". If urban forests are well-designed and well-managed, they can greatly contribute to environmental sustainability, economic viability and liveability in the cities. They help mitigate climate change and natural disasters, reduce energy costs, poverty and malnutrition and provide ecosystem services and public benefits (Salbitano et al. 2017).

This international agency acknowledges that urban development causes depletion and degradation of natural ecosystems, with the subsequent loss of ecosystem services, and the insufficient resilience against disturbances caused by climate change. Quétier et al. (2007) mention that intervening in the ecosystems, whether voluntarily or not, generates changes in the ecological properties (of urban forests, in this case), which have an impact on their capacity of providing ecosystem services. A relevant number of papers describe and classify the ecosystem services provided by urban forests, grouping them into cultural services, regulation or maintenance services and provisioning services (according to the CICES classification 2018).[4]

Assessing ecosystem services necessarily implies not only analyzing them from the perspective of services supplied by the ecosystems, but also from the perspective of services demanded by the users. Therefore, it is important to assess, on one hand, the capacity of urban forests to provide services to society, and on the other hand, the way in which social actors use, value or manage those services (Martín-López 2013).

This chapter analyzes the regulatory framework, considering the acknowledgment, value and/or protection of the services provided by urban forests, and approaches the social perception of these services in the municipality of Luján (Province of Buenos Aires, Argentina). It is a qualitative research of exploratory, bibliographic, documentary and descriptive research which analyzes academic documents, statistical information, legal sources and official and public material.

---

[3]In the document called "Guidelines for Urban and Peri-urban Forestry", published by the Food and Agriculture Organization of the United Nations, *Urban Forests* are defined as networks or systems which encompass all trees, tree groups and individual trees located in urban and peri-urban areas. It therefore includes forests, trees on the streets, trees in parks and gardens, and trees at street corners. Urban forests are the backbone of green infrastructure that connects urban areas with rural areas and improves the cities' environmental footprint.

[4]With the aim of homogenizing international classifications, a Common International Classification of Ecosystem Services (CICES) has been developed. The international Ecosystem Services nomenclature suggests a hierarchical structure with a set of categories and levels that avoid redundancy and overlapping.

2 The Municipality of Luján

The Municipality of Luján is one of the 135 that constitute the Province of Buenos Aires, in Argentina, with a population of 106,273 inhabitants, more than 70% of who are concentrated in the main city (INDEC 2012). It is located 67 km away from the City of Buenos Aires, connected by the *Autopista Acceso Oeste* (West Branch Highway) and the *Sarmiento* Railway Line, and it is crossed by National Routes 5 and 7, and Provincial Routes 6, 47 and 192. Located on the Undulating *Pampas*, this Municipality is part of the middle basin of the Luján River. The climate is sub-humid temperate, with an annual rainfall average of 950 mm (peaking in spring and autumn). Luján has a diversified economy, with an emphasis on the agricultural-livestock sector and on touristic activity, mainly in the surroundings of the historical center, which encompasses the National Basilica, the Pilgrims Rest, *Enrique Udaondo* Museum Complex, the *Recovas* and the Municipal Museum of Fine Arts.

According to a research work conducted by the National University of Central Buenos Aires Province, the Municipality of Luján presents a wellbeing index of 6.76, ranked 141st among 528 Argentine cities. With the aim of weighting the relative incidence of parks and green spaces in this ranking, they have assessed accessibility, size, landscape value, quality and quantity of facilities and the "atmosphere" resulting from the users' behavior and the degree of overcrowding (Velázquez et al. 2010).

3 Results

3.1 Regulatory Advances

We have analyzed laws, codes, ordinances and decrees[5] which determine the uses, define the limits, indicate conditions, foster actions and identify incentives for the management of public and private urban forests at different levels (national, provincial and municipal).

---

[5]According to the hierarchical order of regulations in Argentine Law, lower rank regulations cannot contradict or infringe what has been established by a higher rank regulation. Above all regulations, lie the National Constitution and the International Treaties of Constitutional Hierarchy (included in Section 75, subsection 22), followed by the Integration Treaties and other international treaties. Afterwards, national laws, decrees issued by the national executive power, ministerial resolutions and resolutions issued by other agencies. Provincial regulations fall within Section 31 of the National Constitution (Provincial constitution, provincial laws, decrees issued by the provincial executive power, resolutions and provisions).

3.1.1 At the National Level

It is noteworthy that "Argentina has a federal system where provinces already existed before the Nation was created and therefore have certain powers they have not yet delegated." The inclusion of Section 41 in the Argentine Constitutional Reform of 1994 has been key for the current topic (Minaverry 2018), as it included the right to a healthy and balanced environment, suitable to meet the human needs of the current and future generations. It urges authorities to preserve the natural and cultural heritage, biological diversity and environmental information and education. Pursuant to this amendment, the National State is charged with issuing regulations on the minimum standards for environmental protection, and the provinces shall issue the complementary standards. For example, in 2002, the National State enacted National Law 25,675 (entitled the "General Environmental Law"), which sets forth the importance of preserving the environment for the current and future generations who wish to live in Argentine territory. Following this way, the Province of Buenos Aires also passed Provincial Law 11,723 on the Environment and Natural Resources. Five years later, the National State issued National Law 26,331 on Minimum Standards for the Environmental Protection of Native Forests, which "establishes the minimum standards of environmental protection for the enrichment, restoration, conservation, use and sustainable management of these forests and of the environmental services that they provide to the society." Two years later (in February 2009), its Regulatory Decree (91/2009) was published in the National Official Bulletin. This law acknowledges that native forests provide services to the society (Section 5).

Regarding planted forests, National Law 13,273 on Forest Promotion was passed in 1948, and it introduced a classification of "protective forests" in its Section 8. They are defined as those forests, which due to their location, could serve—collectively or individually—for national defense; for protecting the soil, roads, coastlines, riversides, lake shores, lagoons, canals, islands, ditches and dams and preventing the erosion of plains and slopes; for protecting and regularizing the water regime; fixings banks and dunes; ensuring public health conditions; preventing mud slides and floods; hosting and protecting flora and fauna species.

We have analyzed national legislation that acknowledges the services offered by the forests and promotes forest activity and native forests protection, but there are no specific regulations related to urban forests.

3.1.2 At the Provincial Level

In August 2017, the Province of Buenos Aires passed Provincial Law 14,888 (recently regulated by Provincial Decree 66/17), which approves land-use planning and it is complementary to National Law 26,331 on Minimum Standards for the Protection of Native Forests. Previously, in 2011, an Incentive Plan for the Creation of Production Forests was established by National Law 12,662, with the main goal of generating, expanding and improving the woodlands in the region. Secondarily,

Resolution 338/10 "Provincial Forestation Program—Mitigating Climate Change," issued by the Provincial Agency for Sustainable Development (OPDS), acknowledges the forest ecosystems as "biomes that are highly valued by the society, acknowledged for regulating air, water cycles and basins, for their role in mitigating climate change and for providing goods such as fibers and food".

The Province of Buenos Aires also enacted Provincial Law 12,704 on Protected Landscapes or Green Spaces of Provincial Interest. Section 3 of its regulatory decree (2314/2011) grants priority to urban or peri-urban areas with forested spaces that play a significant environmental role for the population. Following the same line, Provincial Law 10,907 on Reserves and Natural Monuments protects areas that for general interest reasons—mainly scientific, economic, aesthetic or educational— must be spared from human intervention. Moreover, Provincial Law 14,449 on Promoting the Right to Housing and to a Dignified and Sustainable Habitat defines green spaces as an essential component for the development of new residential areas.

The provincial level also incorporated Provincial Law 12,276, specifically focused on urban trees, and was published at the Official Bulletin of the Province of Buenos Aires on April 5th, 1999. It is noteworthy that this legal regulation was issued after the 1994 constitutional amendment that incorporated Section 41 (which has already been mentioned), but in a context where there is no law on minimum standards or any other national law on urban trees. Thus, Buenos Aires set a precedent, becoming one of the eight Argentine provinces[6] to enact an exclusive law on urban tree care, setting its own guidelines regarding this environmental regulation.

Provincial Law 12,276 states that municipalities are accountable for the planning and management of urban trees. It does not mention the services they provide, but its Section 11 grants the possibility of declaring those trees or groups of trees of public interest, thanks to their historical, natural, cultural or aesthetic value. This law defines urban trees as "all tree and bush species installed in an urban or rural area, either municipal or provincial, located in the Municipality's common lands and aimed at public use, regardless of who implanted them or when it happened."

Regarding public green spaces, where we can find the largest surface of urban forests, it is important to mention the Urban Planning Code (UPC) of the Province of Buenos Aires (Provincial Law 8912/87). Section 8b defines "Public and Free Green Spaces" as the "public sectors where vegetation and landscape prevail and its main function is to be used for community recreational purposes and to contribute to purify the environment." The main goals of this UPC are ensuring environmental preservation and improvement, by means of adequately organizing the activities in the space; banning environmentally degrading actions and redressing the already-produced effects; creating physical and spatial conditions that may allow meeting the community requirements and needs at the lowest economic and social

---

[6]Argentine provinces which, in addition to Buenos Aires, enacted laws on urban trees are the following: Salta (6028/1982), San Luis (5478/2004), San Juan (7556/2005), Mendoza (7874/ 2008), Chaco (7162/2012), Tucumán (8991/2017), and more recently, Santa Fe (13,836/2019).

cost, in terms of housing, industry, trade, recreation, infrastructure, equipment, essential services and environmental quality; preserving the sites and areas of natural, landscape, historical or touristic interest; deploying legal, administrative, economic and financial mechanisms that will provide the municipal government with the means to eliminate the speculative excesses, with the aim of ensuring that the urban renewal and planning process will safeguard the general interests of the community; and fostering and stimulating the creation of community awareness of the vital need to preserve and recover the environmental values.

In short, the Province of Buenos Aires is setting a precedent by issuing both the Law on Urban Trees and the protection laws for specific areas, emphasizing the acknowledgment of the services they provide to society.

3.1.3 At the Municipal Level

We have analyzed the regulations issued by the Honorable County Council of the Municipality of Luján, and we have not found any specific regulation on the ecosystem services provided by urban forests (Ferro and Minaverry 2019). Nevertheless, we have reviewed some legal standards that indirectly regulate them. This regulation is associated to the designation of green areas for specific purposes, to the protection of species in institutional premises and to the duty of foresting the urban center perimeter, country clubs and industrial areas, among others. This wide range of legal regulations is related to the existence of different categories of urban forests, located within the Municipality of Luján (Table 1).

The document "Guidelines about Urban Forestry" published by FAO in 2017, gives great significance to a wide range of urban forests and divides them into the following categories:

Peri-urban forests and woodlands: Forests and woodlands surrounding towns and cities, which can provide goods and services such as wood, fiber, fruits, other non-timber forest products, clean water, recreation and tourism.

- City parks and urban forests (>0.5 hectares, ha): Large municipal or urban parks with a variety of land covers and partly equipped with facilities for leisure and recreation.
- Small parks and gardens with trees (<0.5 ha): Small municipal parks equipped with facilities for leisure/recreation, and private gardens and green spaces.
- Trees ′planted on streets or in public squares: tree lined, small groups of trees and individual trees in squares, parking lots, on streets, etc.
- Other green spaces with trees: Urban agricultural plots of land, sports grounds, vacant lands, lawns, riverbanks, open fields, cemeteries and botanical gardens.

Quinta Cigordia Landscape Forest Reserve was created in 1993 by the enactment of Ordinance 3075, and it was declared as a Protected Area throught Ordinance 6772, following Section 10 of Provincial Law 10,907. A Protected Natural Area is

Table 1 Categories of outstanding urban forests in Luján according to regulations

Categories	Urban forests in Luján	Regulation
Peri-urban forests and woodlands	*Quinta Cigordia* landscape forest reserve	• Declared "Environmental Protection District" by Luján's UPC—2019 • Municipal Natural Reserve created by Ordinance 3075/1993 • Declared "Protected Natural Area" by Ordinance 6772/2017 • "Seminar on Heritage Awareness," declared of municipal interest by Decree 754/2019
City parks and urban forests (>0.5 ha)	San Martín park	• Declared "Municipal Natural Heritage" by Ordinance 3763/1997 and categorized as Public Green Space by the UPC • Ordinance 6059/2012 regulates an area which is granted to an Association, with environmental protection guidelines
Pocket parks and gardens with trees (<0.5 ha)	Premises of water supply service	• Declared "Natural Heritage of Forest and Landscape Interest" by Decree 176/2007
Trees on streets or in public squares	Tree lined	• Decree 680 about the expansion of urban centers • Ordinance 2325/88 (modified by ordinance 4294/2001) about the exercise of public woodland
Other green spaces with trees	Banks of the Luján River and Ameghino Park	• Decree, Provincial Law 8912 • UPC, Luján 2019, Environmental Protection District • Ordinance 6051 from May 10th, 2012, where the Charter of the Luján River Basin Committee was signed

Compiled by the authors, adapted from Salbitano et al. (2017)

defined as an area that hosts representative samples of such ecosystem provide environmental services and receives especial protection. The Reserve has an area of 150,000 m^2. It is a woodland which mainly includes exotic species such as privet (*Ligustrum*) and black acacia. It also has interpretive trails and infrastructure (Di Franco 2004). Moreover, Decree 754/19 declared the "Seminar on Natural and Paleontological Heritage Awareness" of municipal interest, held in *Quinta Cigordia* Municipal Natural Reserve in May 2019, which valued its services, considering that "the existence of the Reserve is beneficial to the city, favoring its connection with nature and contributing to the stability of the urban ecosystem, emphasizing the environmental services provided by these areas. Likewise, it is important to point out that its nearness to the touristic, and historical center gives special opportunities for environmental education and interpretation of nature in the Reserve."

Luján's 2019 UPC declares *Quinta Cigordia* Landscape Forest Reserve as an Environmental Protection District, which was defined as an area of special interest in the long term for reassuring sustainability or keeping the settlement's environmental conditions. The green areas, the floodable coasts of the Luján River and of tributary streams of this basin are also considered as such and are identified as areas that should be given special environmental treatment, which implies setting use restrictions and showing interest in recovering them as public spaces. The Urban Reserve represents 37% of the department's surface of green areas, but it presents low-quality and poor accessibility levels (Anselmo et al. 2017).

Within the framework of a project called "Participative and Sustainable Design of Urban Trees in Luján," managed by the Technology Department of the National University of Luján (UNLu), a survey was conducted in 2014 on the wooded area of the Ameghino Park, which is located next to the Luján River and a few meters away from the city center. Its objective was to contribute information about the urban forest status, according to Section 7 of Provincial Law 12,276. Thus, 139 trees were surveyed, predominantly exotic species of the coniferous family. In this place, the park's trees present root and trunk rot, mainly due to the frequent and severe flooding, and it represents a conflicting service, considering the high likelihood of damages caused by tree falls.

On the other hand, Luján's 2019 UPC describes Public Green spaces as large areas in the city intended for public use, for which special parameters are set. It emphasizes that they play a social, recreational, sportive and cultural role, also contributing to preserve the environment.

Point 4.7 of Luján's UPC—about the environmental conditions for urbanization—mentions that it is the duty of urbanization's owner or sponsor to fill the streets with trees and carry out the improvement and landscaping works for the green spaces. Regarding maintenance conditions, it mentions that the Application Authority (AA) is the one in charge of establishing the list of species and recommended plantation distances, considering the features of each species and the municipal regulating plan.

Moreover, item 7.3—about how to register assets with patrimonial value—mentions the forestry and other environmental assets in the following way: "The AA shall be charged with conducting an environmental assets record. This catalogue will register the existing tree species on the streets, in parks and other public spaces. Likewise, it will register the existing species in private plots identifying the different protection levels according to the importance of each species."

We have previously listed some of the AA's responsibilities in the municipality. However, the municipalities have scarce tools to manage green urban infrastructure (Craig et al. 2018), and apparently, Luján is not the exception.

Ordinance 3763 of 1998 declares the San Martín Park as a Municipal Natural Patrimony, but it does not mention the implications of such category. It only forbids the creation of new streets and the circulation of motor-powered vehicles.

Following Ordinance 6059/2012, the Association of Western Children's Soccer Clubs is based in San Martín Park. Covering 5000 m^2 of the public park, the association is committed to maintaining and conserving the granted land, including the trees species, abstaining from withdrawing trees without an express permission of the Executive Department. The ordinance specifies that the association is

Fig. 1 Photograph of San Martín Park (I) taken on October 2nd, 2018. *Source* Analía Scarselletta

committed to identify the existence of dry and/or dangerous trees and to replace them after consulting with the corresponding subject-matter experts.

Within the framework of a seminar entitled "Trees in the Urban Context—2014" organized by the Technology Department of the UNLu, students made a tree census in the Park. It shows scarce diversity of woody species (*Platanus acerifolia* and *Casuarina cunninghamiana* prevail). Residents were also polled, and they mostly expressed a high level of satisfaction about the Park's availability, despite some concerns about insecurity due to the lack of night lighting and to the carelessness of the facilities (Figs. 1 and 2).

> "Many people from Luján use this park to go for a walk, mainly during the weekend, when a lot of people visit the Free Fair…" "I always buy there, they have fresh products, and they value the space, always leaving it clean. It is a nice place, especially in the summer." (González Elizalde, personal communication, October 2nd, 2018)

The Fair works as a trading instance for social and solidarity economy, family agriculture. It seems to be the only instance where the Park is appropriated as a public space.

About another category of urban forest, Decree 176 of 2007, with no local precedent, protects the tree specimens of a land where water supply service premises are located having declared it as "Natural Heritage of Forest and Landscape Interest."

Fig. 2 Photograph of the center of the San Martín Park (II), taken on October 2nd, 2018. *Source* Analía Scarselletta

Regarding tree lined, Decree 680—about the expansion of urban centers—adds onto the above mentioned UPC and states that "urbanized centers shall be forested as follows: one tree for every 12-m façade, two trees for every >12-m and ≤ 20-m façade and one tree every 10 m and/or its fraction for façades of more than 20 m. The fixed number of thirty trees shall be planted in the lands granted for public use. The species must be native and can be chosen from an urban forest species list provided by the application authority, which includes the species varieties and sizes, except those for the lands granted for public use, which will be determined by the Green Spaces Management Office or by the person who holds decision-making power, and they can include up to three varieties chosen from the same list. In all cases, the location of the species to be planted shall be determined by the Green Spaces Management Office or by the person who holds decision-making power."

With the aim of highlighting the value of the supply service offered by urban forests and improving people's quality of life, Ordinance 5997 was enacted in 2011. This regulation suggests benefiting from the urban trees pruning waste, considering which material to choose for the different uses (heating, cooking, composting) and its distribution mechanisms. However, this plan has never been implemented.

Despite all the mentioned regulations, the services provided by Luján's urban forests are not fully contemplated. It is noteworthy that the goods and services

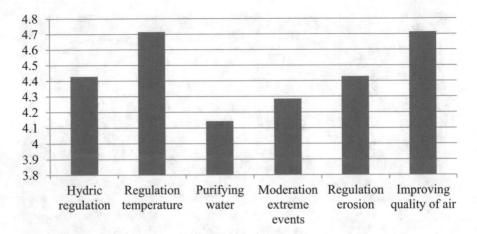

Fig. 3 Average score for each of the regulation or sustainability services offered by urban forests. (1 means "It does not contribute to the service" and 5 means "It successfully contributes to the service"). Key informants' responses; surveys conducted during 2019. Compiled by the authors

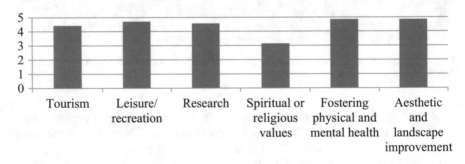

Fig. 4 Average score for each of the cultural services offered by urban forests. (1 means "It does not contribute to the service" and 5 "It successfully contributes to the service"). Key informants' responses, surveys conducted during 2019. Compiled by the author

obtained from the ecosystems bring benefits to societies that largely surpass individual benefits (Capaldo and Minaverry 2016).

3.2 An Approach to Social Perception

With the aim of reviewing the social perception of ecosystem services provided by urban forests, during 2018 and 2019, population-based surveys were carried out, and interviews were conducted with 62 key informants from the Province of Buenos Aires (public managers, university professors, researchers, extension technicians and social organizations referents).

The most valued services have been found to be those of regulation or sustainability, followed by harnessing, and lastly, the cultural services (Figs. 3 and 4).

The most representative supply service considered by key informants was wood supply followed by raw materials for woodworking.

The ecosystem services perceived by Luján residents surpass the services valued by the laws, and the most-frequently mentioned ones were those related to environmental regulation or sustainability. In contrast, in a communiqué addressing the residents' dissatisfaction with the condition of Colón Park, located in the city center, certain antagonistic or conflicting services provided by the trees were mentioned, namely block out of night lighting and visual obstruction due to the tree trunks (Communiqué 87/2001—available on Luján's HCD Digest). The key informants mentioned other perceived conflicting services, such as the risk of falling trees, breaking sidewalks, cars and houses, interference with public services (electricity and telephone wires), allergies, water drainage obstruction and pruning accumulation in the streets. In relation to electricity service, Luján has all the city with overhead power lines which are maintained by the electric cooperative that priorized power lines over lined trees (Fig. 5).

We have gained an approach to the social perception of the services provided by urban forests. Considering both the diversity of the ecosystems and of the population that enjoy them, it would be appropriate to further analyze each of them. On the other hand, a comprehensive assessment should include all three value dimensions: biophysical or ecological, sociocultural and monetary (De Groot et al. 2010).

Moreover, in a survey conducted in 2018 to the population of the Province of Buenos Aires, it was found that 55% of the 335 people interviewed chose the application of strict and high standards (including the application of sanctions) as the most effective measure to solve the environmental problems. The other half considered the implementation of environmental education programs (31%); tax incentives (8%) and delegating responsibilities to non-governmental organizations (6%). In this context, it is appropriate to mention that none of the legal regulations on ecosystem services has a regime of criminal sanctions in their regulatory texts (Minaverry 2016).

4 Final Remarks

According to FAO (2016), municipalities generally regulate urban forests with a guideline resulting from legal regulations that are enacted and amended over time. Nevertheless, those regulations could reinforce local communities' capacity to control and manage the trees located both in public and private lands. According to the Argentine institutional structure, municipal autonomy–understood as the possibility of local communities to govern themselves free from any other State interference–does not represent an absolute concept, but rather has different degrees and kinds (Minaverry 2013). This makes it more complex to adapt to national and

Fig. 5 Word cloud resulting from the population-based surveys, following the question: Do you think trees in an urban context offer services? Could you mention any? Compiled by the authors

provincial regulations to be applied at the municipal level. However, as it has been acknowledged by Ferro and Minaverry (2019), it is worth stressing the contribution made by Luján's ordinances, which have incorporated the ecosystem approach guidelines more than provincial and national jurisdictions.

To conclude, we believe that although there are new regulations that consider different forms of urban and peri-urban forests and acknowledges the ecosystem services they provide, it is necessary to offer guidance, incentives, and to specifically state rights and responsibilities. All the foregoing aims to strength and maintain those ecosystems, so that they can meet the necessary conditions and contribute to improve quality of life both for the present local and global population and for the future generations.

References

Anselmo J, Di Franco L, Cucciufo E, Craig E (2017) Metodología para el relevamiento de los espacios verdes públicos en la ciudad de Luján, provincia de Buenos Aires. Geografía y Sistemas De Información Geográfica (GEOSIG) 9(9):34–52

Craig E, Cucciufo E, Di Franco L, Bormioli N, Esquivel J (2018) Methodological proposals for risk evaluation of urban forest in the metropolitan area of Buenos Aires, Argentina. In: World forum on urban forests. Mantova

De Groot RS, Alkemade R, Braat L, Hein L, Willemen L (2010) Challenges in integrating the concept of ecosystem services and values in landscape planning, management and decision making. Ecol Complex 7:260–272

Di Franco L (2004) Relevamiento de especies forestales a través de fotografías aéreas. Universidad Nacional de Luján. Laboratorio de PRODITEL, Buenos Aires

INDEC (2012) Censo nacional de población, hogares y viviendas 2010. Censo del Bicentenario. https://www.indec.gob.ar/. Last access 30 Nov 2019

Ferro M, Minaverry CM (2019) Aportes normativos, institucionales y sociales a la gestión del agua y el enfoque ecosistémico en la Cuenca del Río Luján, Argentina. Revista De Derecho 20:25–55. https://doi.org/10.22235/rd.vi20.1859

IPBES (2018) The IPBES regional assessment report on biodiversity and ecosystem services for the Americas. In: Rice J, Seixas CS, Zaccagnini ME, Bedoya-Gaitán M, Valderrama N (eds) Secretariat of the intergovernmental science-policy platform on biodiversity and ecosystem services. Bonn, Germany. Last access 8 Aug 2019. https://doi.org/10.5281/zenodo.3236252

Martín-López B (2013) Evaluación de los servicios de ecosistemas suministrados por cuencas hidrográficas: una aproximación socioecológica. Universidad Autónoma de Madrid, Madrid

Minaverry CM (2013) La protección legal del agua potable en Argentina y su inclusión en la agenda internacional. In: Book of presentation of XI Congreso Nacional de Ciencia Política. Sociedad Argentina de Análisis Político y Universidad Nacional de Paraná, Entre Ríos, pp 1–34

Minaverry CM (2016) Consideraciones sobre la regulación jurídica ambiental de los servicios ecosistémicos en Argentina. Estudios Sociales, Centro De Investigación En Alimentación y Desarrollo 26(48):43–66

Minaverry CM (2018) El derecho ambiental en la gestión de los bosques nativos (Espinal) en Argentina. Revista Sociedad y Ambiente 16:157–177

Quétier F, Tapella E, Conti G, Cáceres D, Díaz S (2007) Servicios ecosistémicos y actores sociales. Aspectos conceptuales y metodológicos para un estudio interdisciplinario. Gaceta Ecológica 84(85):17–26

Salbitano F, Borelli S, Conigliaro M, Chen Y (2017) Directrices para la silvicultura urbana y periurbana. Estudio FAO, Roma. http://www.fao.org/3/b-i6210s.pdf. Last access 28 June 2019

Velázquez G, Mikkelsen C, Linares S (2010) Calidad de vida en Argentina: ranking del bienestar por departamentos. Universidad Nacional del Centro de la Provincia de Buenos Aires, Buenos Aires

International Instruments

United Nations (2016) New urban agenda. In: United Nations conference on housing and sustainable urban development (Habitat III). Held in Quito, Ecuador, on October 20th, 2016. ISBN: 978–92–1–132736–6

United Nations (2018) 2030 Agenda and sustainable development goals: an opportunity for Latin America and the Caribbean (LC/G.2681-P/Rev.3)

National Regulations

Argentina. National Law 25,675, November 2002
Argentina. National Law 26,331, December 2007
Argentina. Regulatory Decree 91, February 2009
Argentina. Buenos Aires Provincial Law 13,273. November 1995

Provincial Regulations

Buenos Aires. Provincial Law 11,723, December 1995
Buenos Aires. Provincial Law 14,449. November 2012
Buenos Aires. Provincial Law 14,888. December 2016
Buenos Aires. Provincial Law 12,662. March 2001

Buenos Aires. Provincial Law 12,704. May 2001
Buenos Aires. Provincial Law 10,907. April 1990
Buenos Aires. Provincial Law 12,276. March 1999
Buenos Aires. Regulatory Decree 366, Provincial Law 14,888. June 2017
Buenos Aires. OPDS (Provincial Agency for Sustainable Development). Resolution No. 338.
    November 2010 "Provincial Forestation Program—Mitigating Climate Change", Province of
    Buenos Aires
Buenos Aires. Decree, Provincial Law 8912, August 4th, 1987, La Plata

Municipal Regulations

Municipality of Luján (1993). Ordinance 3075. Province of Buenos Aires
Municipality of Luján (1998). Municipal Ordinance 3763. Province of Buenos Aires
Municipality of Luján (2007). Decree 176. Province of Buenos Aires
Municipality of Luján (2012). Ordinance 6059. Province of Buenos Aires
Municipality of Luján (2012). Ordinance 6051. Province of Buenos Aires
Municipality of Luján (2017). Ordinance 6772. Province of Buenos Aires
Municipality of Luján (1988). Ordinance 2325. Province of Buenos Aires
Municipality of Luján (2019). UPC—Luján. Province of Buenos Aires
Municipality of Luján (2019). Decree 680. Province of Buenos Aires
Municipality of Luján (2019). Decree 754. Province of Buenos Aires

Introduction—Indigenous Presence and Current Legislation in Argentina

Juan Manuel Engelman and Sofía Micaela Varisco

Abstract This chapter provides general information and background about indigenous people presence in Argentina and about the main valid laws about this issue. Argentina is located in the southern portion of the American continent, and its territory consists of six main, large regions: Patagonia in the South; the *Pampas* in the Center; Cuyo, the western Andean mountains; the Autonomous City of Buenos Aires (CABA) and the Greater Buenos Aires (GBA); and the Northwestern and Northeastern regions. All these areas are currently the home place of more than 40 indigenous peoples, amounting to approximately one million inhabitants (out of a total population of 45 million people in Argentina). In demographic terms, the largest groups are the Mapuche people (with a population of over 200,000), followed by the Toba (Qom) and the Guaraní (with less than 200,000 and more than 100,000), and finally, the Diaguita, Kolla, Quechua and Wichí groups, encompassing a population between 50,000 and 100,000 (INDEC in 2010 National Population and Housing Census—bicentenary census. Indigenous peoples, metropolitan region. Series D. Number 6. Buenos Aires, 2015). Therefore, 70% of Argentina's indigenous population is concentrated in seven major groups, while the remaining 30% is fragmented into more than 30 groups of lower demographic significance.

Keywords Argentina · Indigenous Peoples · Laws · Presence

J. M. Engelman (✉)
Consejo Nacional de Investigaciones Científicas y Tecnológicas Facultad de Filosofía y Letras, Universidad de Buenos Aires - Universidad Nacional de Luján, Argentina, Buenos Aires, Argentina

S. M. Varisco
Fondo para la Investigación Científica y Tecnológica, Programa de Arqueología Histórica y Estudios Pluridisciplinarios, Universidad Nacional de Luján, Luján, Argentina

1   Introduction

Argentina is a country located in the Southern portion of the American continent. It comprises six regions: Patagonia in the South; the *Pampas* in the Center; Cuyo, the western Andes mountains; the Autonomous City of Buenos Aires (CABA) and the Greater Buenos Aires (GBA); and the Northwestern and Northeastern regions. All these areas are home to more than 40 indigenous peoples, whose settlements bring together a population of approximately one million inhabitants, out of a total population of 45 million people countrywide.

Although the indigenous presence in our country is not as demographically significant as it is in other countries of Latin America—where it matched the State civilizations, such as the Andean and Mesoamerican cultures—the indigenous peoples are an essential part of our origins and of the current reality in many local areas. Such is the case of some departments in the provinces of Salta, Formosa, Jujuy, Neuquén, Río Negro, Chubut, and some specific districts throughout the country, including the province of Buenos Aires.

According to the data provided by the 2010 National Population and Housing Census, the demographically largest groups are the Mapuche people (with more than 200,000 members), the Toba (Qom) and Guaraní peoples (with less than 200,000 and more than 100,000), followed by the Diaguita, Colla, Quechua and Wichí groups, with 50,000–100,000 members. Therefore, 70% of Argentina's indigenous population is part of these seven major groups, while the remaining 30% belongs to more than 30[1] groups (INDEC 2010).

Unlike other Latin American countries, one of the characteristics of the Argentine Nation-State during its creation period was the predominance of a positivist ideology that embedded the myth of a nation with European roots in the collective imagination. This explains why, historically, there was a tendency to deny indigenous presence until well into the twentieth century. In addition, it also explains why the policies addressing indigenous peoples are relatively recent (as we will explain later).

Currently, the presence, organization and visibility of the indigenous peoples have become an input to counteract the widespread idea that they exclusively live in rural areas—an idea based on the hegemonic discourse that hides and denies them. It is worth mentioning that various organizational experiences have also been taking place jointly between rural and urban areas.

According to the census (INDEC 2010), the provinces with the largest proportion of native population (over 5% of the total population) are Chubut, Neuquén, Jujuy, Río Negro, Salta, and Formosa. However, it is important to point out that indigenous population is often underrepresented in statistical data and that their actual presence is larger (Trinchero 2010). This can occur either because they were

---

[1]Comechingón (3.6%), Huarpe (3.6%), Tehuelche (2.9%), Mocoví (2.3%), Pampa (2.3%), Aymara (2.2%), Ava Guaraní (1.9%), Rankulche (1.6%), Charrúa (1.5%), Atacama (1.5%), Mbyá Guaraní (0.8%), Omaguaca (0.7%), Others (4.3%) (INDEC 2010).

not recognized, or because the censuses were poorly implemented, or because of an extremely widespread prejudice that considers "being indigenous" as a negative condition. Furthermore, in some cases, indigenous people are not included in the census if they are living "in the cities," as it is wrongly assumed that "their original place should be the rural one" (Engelman 2019).

As mentioned above, indigenous presence in Argentina has been historically denied, and therefore, policies focusing on indigenous peoples are relatively recent. Historically, the State's attitude toward these groups was initially belligerent (from the nineteenth century until well into the twentieth century), and later it tended not to visibilize them.

With the return of democracy in 1983, and especially over these last few years, indigenous peoples have gained their recognition as subjects of law, thanks to their struggles and various negotiation processes with the State. This allowed the enactment of different laws aimed at making their indigenous identity and territorial claims effective. The pre-existence of the indigenous population before the creation of the Argentine State was acknowledged, providing a legal corpus and a basis for the subsequent ethnic and political organization of these populations.

In 1992, Argentina signed the Convention N° 169 of the International Labor Organization (ILO)[2]—ratified in 2001 by National Law 24,071—which entailed a major shift away from the integrationist notion that proposed to assimilate indigenous communities within the official culture. Currently, this convention is one of the most important legal instruments addressing the rights of indigenous peoples at an international level.

Moreover, as the indigenous peoples increasingly obtained recognition for their claims, and as a result of different problems that became more complex for the indigenous peoples over the years (such as those linked to deforestation, and privatization and/or transfer of their lands to foreign ownership), different laws have been enacted. These regulations focus on the indigenous peoples or on broader social sectors, but they have a particular effect among the indigenous groups. We refer to those laws that regulate the land and/or environmental aspects, such as National Law 26,160 on "Declaration of Emergency in the Matter of Possession and Ownership of Land traditionally occupied by Indigenous Communities,"[3] National Law 26,737 ("Protection of National Control of Property, Possession or Tenure of Lands Law," colloquially known as the "Land Law") and the Law which is the main focus for the purpose of this chapter—National Law 26,331 on "Minimum Standards for the Environmental Protection of Native Forests," also known as the "Forest Law" which was passed in 2007.

---

[2]The International Labour Organization (ILO) was created in 1919, and it is a specialized agency of the United Nations (UN) which focuses on establishing regulations aimed at improving the living and working conditions in the whole world. The ILO has adopted several conventions on the protection of workers, many of which are applicable to indigenous and tribal peoples.

[3]On the one hand, it acknowledges the precarious land situation, and at the same time, it prevents evictions (while the emergency lasts). On the other hand, it establishes that studies must be carried out to survey the situation of the different communities.

The emergency background of National Law 26,331 is the inclusion of environmental rights after the amendment of National Constitution in 1994. In its Section 41, it guarantees the right of all inhabitants to enjoy a healthy environment, and it establishes the division of responsibility, determining that the nation is responsible for "enforcing the regulations included in the minimum protection standards, and the provinces are responsible for any necessary complementary regulation, without altering the local jurisdictions." In other words, although the provinces can exercise their original domain of the natural resources in their territories (Section 124), the exercise of this authority must follow certain minimum criteria of environmental protection duly established by the nation (Schmidt 2018).

It is important to clarify that the implementation of the different laws faces great difficulties, especially those related to land regulation. This is the case of National Law 26,160 on "Indigenous Land Emergency," which was passed in 2006 (and had to be extended in 2009, 2013 and 2017 due to delays in its implementation) (Varisco et al. 2018). The last extension was achieved in November 2017, amidst a context of great indigenous mobilization, increasing conflict and stigmatization of the different peoples, especially the Mapuche people (in the face of various interests that sought to overrule this Law). In fact, 13 years after its enactment, only about half of the 1687 communities identified by the National Program for Territorial Survey of Indigenous Communities (RETECI)—which belongs to the National Institute of Indigenous Affairs (INAI)—have been surveyed, and in only 38% of the cases the entire process has been completed and the community has received a termination resolution (Amnesty International 2019) (Fig. 1).

Nowadays, at least 720 registered indigenous communities throughout the country have not been surveyed. Furthermore, many of them are facing great conflict, are strongly affected by tourism, by real estate activity and by agribusiness and deforestation.

The fact that what is mandated by National Law 26,160 has not been completed or implemented creates an impact on other regulations—such as the Forest Law—since it prevents communities from being acknowledged as subjects of law and from accessing to benefits (such as financing various projects related to this Law).

This chapter described an introduction to the indigenous peoples in Argentina and to the different laws that have been implemented in that regard, which have an impact on them. The following chapters, written by different authors, will address the Forest Law's implementation and the specific problems it presents concerning indigenous peoples, focusing on the Chaco and Northern Patagonia regions (Fig. 2).

Fig. 1 Mobilization in Congress, September 2017. *Source* Programa Entnicidades y territorios en redefinición -Instituto de CienciasAntropológicas, Facultad de Filosofía y Letras, Universidad de Buenos Aires

Fig. 2 First meeting in defense of the Intercultural Education of music and painting (year 2012). School 357, Mapuche Painefilú community, Junín de los Andes, Neuquén province. *Source* Programa Entnicidades y territorios en redefinición -Instituto de CienciasAntropológicas, Facultad de Filosofía y Letras, Universidad de Buenos Aires

References

Amnistía Internacional (2019) Estado de situación de la ley de Emergencia territorial indígena 26160. https://amnistia.org.ar/wp-content/uploads/delightful-downloads/2019/10/Informe_EstadodeSituacion20160_ok.pdf. Last access 15 Oct 2020

Castilla M, Varisco S, Valverde S (2018) Políticas de Intervención con los Pueblos originarios Mapuche y Qom en Argentina. Revista De Estudos e Pesquisas Sobre as Américas 12(2):86–123

INDEC A (2010) National population and housing census. Buenos Aires, Argentina

INDEC (2015) 2010 national population and housing census—Bicentenary census. Indigenous Peoples, Metropolitan Region. Series D. Number 6. Buenos Aires

Engelman JM (2019) Indígenas en la ciudad: articulación, estrategias y organización etnopolítica en la Región Metropolitana de Buenos Aires. QUID 16(11):86–108

Schmidt M (2018) Una década protegiendo los bosques nativos. Claroscuros de una política ambiental en defensa de los bienes comunes. Revista Bordes de Política, Derecho y Sociedad s/n:153–162

Trinchero H (2010) Los Pueblos Originarios en la formación de la Nación Argentina. Revista Espacios 46:106–123

Forests, Culture and Health

Mariana Costaguta, Martín Rodríguez Morcelle, Laura Gabucci, and Bruno Lus

Abstract The territory is a multicultural space with traditional and/or circumstantial and recent knowledges and practices, which presents a complexity and dynamics which enriches itself. Biological and cultural diversity nourish with migrant population exchanges, which are generations which provide benefits with popular health construction. The concept of health is a dynamic construction. Nevertheless, tradition knowledge is linked to oral and intergenerational transmission, which relates to the spiritual and depends on the access to sacred spaces and/or territories where there are biological and symbolic resources, conforming a holistic paradigm. There are more than 40 native groups of people living in Argentina, which are distributed in different phytogeographic regions which coexist in a relationship of respect for nature. One of the most sustainable exploited natural resources is the large quantity of plants used for health issues. The dialogue between perspectives and paradigms requires to avoid translations, implementing wrong interpretations which denaturalize these knowledge and practices. The identification, characterization and monitoring of ecosystem services with health benefits would provide better access and reproduction of culture of people. It is fundamental to promote public policies which guarantee cultural reproduction in a framework of ecosystem health.

Keywords Cultural ecosystem services · Benefits · Population

1 Introduction

According to FAO (2012), a forest is a tract of land of more than 0.5 ha, with trees more than 5 m high and a canopy cover greater than 10%, or with trees that can reach this height in situ. This definition is biased and quite inaccurate. On the one

M. Costaguta (✉)
Vice Coordinator, TRAMIL, Buenos Aires, Argentina

M. R. Morcelle · L. Gabucci · B. Lus
Universidad Nacional de Luján, Luján, Argentina

© The Author(s), under exclusive license to Springer Nature Switzerland AG 2021
C. M. Minaverry and S. Valverde (eds.), *Ecosystem and Cultural Services*, The Latin American Studies Book Series, https://doi.org/10.1007/978-3-030-78378-5_5

hand, forests are not acknowledged as complex ecosystems, ignoring other plant components (herbs, shrubs and vines), animals and organisms from other kingdoms, such as fungi. Likewise, it is essential to mention the matter and energy interrelations that occur between biotic and abiotic factors in a forest. Moreover, this definition does not consider human communities that inhabit those forests. However, in 2018, FAO acknowledged that forests represent a source of food, medicine, and fuel for over one billion people worldwide (FAO 2018). This perspective stems from the paradigm on which capitalism is based, which has an anthropocentric and extractivist viewpoint. On the other hand, in our continent, several native populations consider themselves part of this ecosystem given their worldview. The forest is seen as their livelihood and not only as a resource.

Forests are estimated to host 65% of the planet's biological diversity. They are an important and direct source of food, medicine, fuel and construction elements for human beings. They also play an important indirect role in preserving water sources, avoiding floods and protecting the soil and are responsible for holding a large amount of carbon dioxide, one of the main greenhouse gases. Thus, approximately 1.6 billion people depend on forests to obtain food, and around 60 million indigenous people from different regions exclusively depend on these areas for their livelihoods (Fundación Vida Silvestre Argentina—*Argentine Wildlife Foundation* 2020).

Argentina has a large stretch of woodlands and forests, which Cabrera (1971) calls the phytogeographic provinces: the Paraná, Yungas, Chaco, Espinal and Patagonia areas. Let us focus on the case of the Patagonian Andean forests. They expand along a narrow strip of less than 70 km along the 2,200 km of the Andes Mountains, from Neuquén to Tierra del Fuego in Argentina, and from Maule to Magallanes in Chile (Barthelemy et al. 2008). They cover an area of 6,446,523 ha, and they typically grow in predominantly mountainous environments, in a cold and humid climate, favored by the heavy rainfall caused by humid air masses coming from the Pacific Ocean (Proyecto Bosques Nativos y Areas protegidas 2003). Therefore, they have become one of the planet's most important mountain latitudinal biological corridors (Goya et al. 2005). They have a great flora diversity, where important species stand out, namely: trees of the *Nothofagus* Blume genus (lenga, coihue, raulí, Patagonian oak, ñire and cherry), mountain cypress (*Austrocedrus chilensis* (D. Don) Pic. Serm. and Bizzarri), pehuén (*Araucaria araucana* (Molina) K. Koch), larch (*Fitzroya cupressoides* (Molina) I.M. Johnst.), radal (*Lomatia hirsuta* (Lam.) Diels), maitén (*Maytenus boaria* Molina), as well as other emblematic species such as the colihue cane (*Chusquea culeou* E. Desv.) and maqui (*Aristotelia chilensis* (Molina) Stuntz), among others (Barthelemy et al. 2008; Demaio et al. 2017). All of them have been used by humans for different purposes, as food resources, medicine, shelter, construction, and decoration, among others (Demaio et al. 2017).

Evidence shows that Patagonia has been inhabited at least for 13,000 years and that in the pre-Columbian period approximately 300,000 inhabitants came to occupy these lands. The Patagonian provinces are characterized by currently having a high percentage of indigenous inhabitants or descendants, surpassing the national average. In the province of Chubut, for example, 8.7% of households identify themselves as part of this group, when the national average is 2.4%. In Argentina, almost one million people identify themselves as part of or as descendants of native peoples—the Mapuche people, originally from the Patagonian region, being the largest with more than 200,000 people (INDEC 2010).

Forest ecosystems are the source of several plants used for health care. In the Argentine Patagonian region, different plant species are used for medicinal purposes. Molares and Ladio (2009) surveyed a total of 505 species categorized as medicinal, 60% of which were native and 40% exotic. The categories where they are used include gastrointestinal, dermatological, genitourinary, circulatory and heart, respiratory, for pain and inflammation, fever, and related to cultural syndrome, among others. For example, the endemic native bush[1] called "pañil," "matico" or "palguñi" (*Buddleja globose*, Hope) is used both as an infusion for digestive ulcers and hepatic dysfunction and as an external compress to wash the skin. It has analgesic, healing, antifungal and anti-inflammatory properties (Kutschker et al. 2002). Western science, through pharmacology, attributes these properties to the presence of secondary metabolites like iridoids and flavonoids.

Criollo and Mapuche communities use it based on their ancestral knowledge, passed on from generation to generation and also did its effectiveness. A native species of Europe and Asia, naturalized in the region and frequently used, is the "pilunhueque," "trafue" or "llantén" (*Plantago lanceolata* L.) which is used for stomachaches, as oral and respiratory tract anti-inflammatory and for eye cleaning. Another species called "llantén" (*Plantago major* L.), also naturalized in the region, is used for earaches, gastritis, ulcers and bronchial colds (Itkin 2004). In the Caribbean basin, the same species is also used for medicinal purposes, but for nervous breakdowns and stress (Germosén-Robineau 2014). There is a significant percentage of exotic medicinal species present in the Mapuche Pharmacopoeia, and this phenomenon also occurs in indigenous communities from other regions of our country. This fact reflects these peoples' ability to incorporate new knowledge (the use of plants, in this case) beyond cultural imposition and the processes aiming to marginalize and exclude native communities.

However, it is important to try and identify what a plant's healing capacity is based on. Very often, it is not possible to explain this in pharmacological terms, and it can even happen that the same plant is not indicated for two different people with the same health problem. For example, in the case of the Mapuche people, each

---

[1]Endemic: a plant species is endemic when its geographic distribution is restricted to a single place.

medicinal plant has an owner (*lawen*), who gives the Mapuche specialist (*machi*) the possibility to use it for therapeutic purposes. Therefore, it is not that the plant has a healing power, but that its supernatural origin allows for the restoration of the sick person's balance. Molares and Ladio (2009) pointed out that "most of those (exotic)[2] species appeared at the area approximately 300 years ago, with the arrival of European settlers, long enough for them to have been adopted, named in the vernacular language, cultivated and valued as significantly important healing elements" (p. 109).

2 Culture and Territory

Forests therefore develop in different climates, reliefs and latitudes, spreading over specific territories and becoming the livelihoods for their local communities.

In the case of our country, an indigenous group is a group of families and communities identified with a common history that precedes the creation of the Argentine Nation. They have their own culture and social organization, and they are linked by a distinctive language and identity. Having shared a common territory, they currently keep part of it through their communities (Ministry of Culture 2018).

The territory then becomes a multicultural space where traditional and/or recent and circumstantial knowledge and practices coexist. It is not necessarily a map in the cartographic sense, but according to Sánchez Ayala (2015), it could be interpreted as "...a spatial entity that serves as a communication instrument, making social structures visible and tangible—such as authority, identity, rights, aspirations and prejudices, among many others" (p. 176). It could be considered as a dimension characterized by a boundary with fuzzy edges. In this case, a forest would be part of a territory both inhabited by one or more cultures and related to an "outside," thus embodying an enriching complexity (Sánchez Ayala 2015).

Likewise, it is important to remember that the analysis of any type of intervention on a "forest" should include the strategies, regulations and public policy controls, as they are locally managed at different levels within the various State levels (national, provincial and municipal). For example, in 2007 Argentina passed the National Law 26,331 on Minimum Standards for the Environmental Protection of Native Forests (also known as the "Forest Law"). Among other points, this law establishes that each province must draft a land-use planning, defining a category type for each existing native forest, based on an attributable conservation level (Alcobé 2020).

On the other hand, traditional knowledge is linked to oral and cross-generational transmission, deeply rooted in the spiritual field, and it depends on the access to sacred places and/or territories, where biological and symbolic resources are found, in a so-called holistic paradigm. Therefore, the environmental quality of these

---

[2]Author's note.

territorial spaces is what ensures that a culture will be reproduced and will continue over time.

2.1 Medicinal Plants as a Cultural Value for Health

Some native peoples in Latin America have a vast plant pharmacopoeia included in their health practices, while others are restricted to just a limited number of species, as they focus on other types of knowledge. Not all peoples have phytotherapy as a healthcare alternative. However, those who do use medicinal plants have their own unique plant types each, as well as their own preparation methods, dosage and diagnosis. The same plant might be used in a specific way in one community and in a different way in another. In order to address the medicinal aspects of a forest's biodiversity, for example, we must explain the structural concepts of the health-disease-care process experienced by the communities. We agree with Menéndez (2009) that "self-care constitutes one of the basic activities of the health-disease-care process, being the nuclear and synthesizing activity developed by social groups and individuals regarding this process. By self-care, we refer to the representations and practices a population uses at individual and/or social group level to diagnose, explain, care for, control, relieve, support, heal, solve or prevent the processes that have an impact on their health in real or imaginary terms, with no direct and intentional intervention of professional healers" (Menéndez 2009, p. 52). Medicinal plants are one of the tools that add meaning to the concept of self-care, since people attribute a preventive or healing function to them, causally related to their vision and culture (Castiel 2006).

Although there are many ethnobotanical works that link plant lists to their applications for certain organic systems such as digestive and nervous, we believe that an effort must be made to link this systematization to a local epidemiological profile with a technical scientific as well as sociocultural basis. Frequently, an epidemiological profile shows the importance of chronic diseases and infectious problems, among others. However, it usually happens that the way health problems or their causes are called differs from what a certain culture may call it. Or that the relative importance attributed to them is different. For example, a community may consider it as important to address high blood pressure as to address fear, root-lessness, violence, transgression to sacred places, envy or other categories embodied by the culture itself, and which health professionals or researchers fail to notice in their transient passage through the community space.

The concept of health is a dynamic construction, which changes over time and according to living conditions and their social determinants. The following questions should always be present when trying to learn about the medicinal use of the local flora: What are the typical illnesses suffered by this community? What do they

Fig. 1 Cohiue forests (*Nothofagus dombeyi*). Lago Puelo National Park, Chubut, Argentina. 2020. *Source* Photograph taken by Bruno Lus

say they suffer from? Which are the main pathologies from a sociocultural epidemiology perspective? How do they cure them? Which are the existing accessibility barriers? Beyond the language and cultural barriers, interdisciplinary and collective work would allow us to approach a process that may have a positive effect on these groups' health care. Somehow, the key might be to become "interpreters" and not "translators." Maybe the local flora is not just a floristic inventory in terms of resources, but a cultural value for health care.

Therefore, within the context of these different perspectives, making an inventory of the medicinal flora through documented herbaria, promoting local cultivation in medicinal gardens or preserving the biodiversity of the ecosystems inhabited by these species could represent concrete tools to foster research and describe the local vademecum at the same time.

In the last decade, the university extension team of Botany at the Universidad Nacional de Luján has developed projects about medicinal plants, focusing on the defense and revaluation of the knowledge, culture and practices of the health-disease-care process, within the university population-health sector and educational interface, and thus directly contributing to the promotion of intercultural

Fig. 2 Phytogeographic province of the Patagonian Andean forests. Lago Puelo National Park, Chubut, Argentina. 2020. *Source* Photograph taken by Bruno Lus

health practices, and consequently, to the indirect promotion of local biodiversity (Costaguta et al. 2014). Our goal is to create a positive impact on health by fostering inclusive public policies with an intercultural perspective and based on evidence of both cultural and scientific use of medicinal plants (Rodríguez Morcelle et al. 2018).

Our analysis is based on our practice in primary healthcare centers and on our ongoing dialogue with the teams of professionals related to those centers.

Thus, a forest is much more than an ecosystem from the biological perspective; it is part of the dynamics of the living organizational forces of a local culture, which could in turn become the protecting and multiplying agent for biodiversity (Figs. 1, 2, 3, 4 and 5).

Fig. 3 Phytogeographic province of the Yunga. Tucumán, Argentina. 2013. *Source* Photograph taken by Bruno Lus

Fig. 4 Phytogeographic province of the Prepuna. Jujuy, Argentina. 2013. *Source* Photograph taken by Bruno Lus

Fig. 5 Phytogeographic Paranaense province. Misiones, Argentina. 2014. *Source* Photograph taken by Bruno Lus

References

Alcobé F (2020) Los Bosques Nativos de Argentina en el marco del proceso de Reducción de Emisiones derivadas de la Deforestación y la Degradación (REDD). Last access: 9 Oct 2020. https://www.undp.org/content/dam/argentina/Publications/Energia%20y%20Desarrollo%20Sostenible/brief-08-cambios.pdf

Barthelemy D, Brion C, Puntieri J (2008) Plantas de la Patagonia. Vázquez Mazzini Editores, Buenos Aires

Cabrera AL (1971) Fitogeografía de la República Argentina, Boletín de la Sociedad Argentina de Botánica, vol XIV. Buenos Aires, pp 1–2

Castiel LD (2006) Dédalos y los dédalos: identidad cultural, subjetividad y los riesgos para la salud. In: Czeresnia D, Machado de Freitas C (eds) Promoción de la Salud. Conceptos, reflexiones y tendencias. Lugar Editorial, Buenos Aires

Costaguta M, Lus B, Gabucci L, Rodríguez Morcelle M (2014) Plantas medicinales: promoción de la salud intercultural comunitaria desde la universidad. Rev Extensión +E 4:74–79

Demaio P, Karlin U, Medina M (2017) Árboles Nativos de Argentina. Tomo 2, Patagonia. Ecoval Editorial, Córdoba, Argentina

FAO (2012) Documento de Trabajo de la Evaluación de los Recursos Forestales No. 180. Last access: 9 Oct 2020. http://www.fao.org/3/ap862s/ap862s00.pdf

FAO (2018) El estado de los bosques del mundo. Las vías forestales hacia el desarrollo sostenible. Last access: 9 Oct 2020. http://www.fao.org/3/I9535ES/i9535es.pdf

Fundación Vida Silvestre Argentina (2020) ¿Cuál es el problema? Los bosques, en peligro. Last access: 9 Oct 2020. https://www.vidasilvestre.org.ar/nuestro_trabajo/que_hacemos/nuestra_solucion/cambiar_forma_vivimos/conducta_responsable/bosques/_cual_es_el_problema_/

Germosén-Robineau L (ed) (2014) Farmacopea Vegetal Caribeña. Tramil, Mexico

Goya JF, Frangi JL, Arturi MF (2005) Ecología y manejo de los bosques de Argentina: Investigación en bosques nativos de Argentina. Last access: 9 Oct 2020. https://libros.unlp.edu.ar/index.php/unlp/catalog/view/438/404/1438-1

INDEC (2010) Censo Nacional de Población, Hogares y Viviendas 2010. Last access: 9 Oct 2020. https://www.indec.gob.ar/indec/web/Nivel4-Tema-2-21-99

Itkin S (2004) Plantas de la Patagonia para la salud. Caleuche. San Carlos de Bariloche, Río Negro

Kutschker A, Menoyo H, Hechem V (2002) Plantas medicinales de uso popular en comunidades del oeste de Chubut. Bavaria, San Carlos de Bariloche, Río Negro

Menéndez E (2009) De sujetos, saberes y estructuras. Introducción al enfoque relacional en el estudio de la salud colectiva. Editorial Lugar, Buenos Aires

Ministerio de Cultura (2018) Los Pueblos Originarios en Argentina, hoy. Last access: 9 Oct 2020. https://www.cultura.gob.ar/dia-internacional-de-los-pueblos-indigenas_6292/

Molares S, Ladio A (2009) Plantas medicinales de los Andes Patagónicos: una revisión cuantitativa. In: Vignale N, Pochettino ML (eds) Avances sobre plantas medicinales andinas. CYTED, San Salvador de Jujuy

Proyecto Bosques Nativos y Áreas Protegidas (2003) Atlas de los Bosques Nativos Argentinos. Dirección de Bosques. Secretaría de Ambiente y Desarrollo Sustentable. Buenos Aires, Argentina

Rodríguez Morcelle M, Gabucci L, Lus B, Costaguta M (2018) Las plantas medicinales en Luján: un punto de encuentro de saberes. In: Cucciufo E, Di Matteo J, Colombo M (Compilators) Abriendo Caminos. EdUNLu, Luján, Buenos Aires

Sánchez Ayala L (2015) De territorios, límites, bordes y fronteras: una conceptualización para abordar conflictos sociales. Revista De Estudios Sociales 53:175–179

Forests and Public Policies in the Argentine Northern Patagonia Region—Small Producers, Capitals, and Territorial Claims

Valeria Iñigo Carrera, Alejandro Balazote, and Gabriel Stecher

Abstract The mountainous area in the provinces of Neuquén and Río Negro (Argentine Northern Patagonia) is extremely rich in terms of native forest cover, soils, grasslands, waters, and landscapes. Consequently, the area is highly valued for tourism and its associated activities (real estate, for example), among others. The increase of national and foreign capitals applied to these activities entails a threat for the traditional dwellers of these lands and territories: small producers who identify themselves as belonging to indigenous peoples, and those with a *criollo* origin. In 2007, National Law No. 26,331 on "Minimum Standards for the Environmental Protection of Native Forests" was enacted, as well as Provincial Laws No. 2,780 and No. 4,552 in Neuquén and Río Negro provinces, respectively. From then on, the management of forest areas has been subject to regulation by both the federal and the provincial states. This chapter examines the implications of public policies regarding nature conservation—in general—and territorial planning of native forests—in particular—in relation to the territorialities configured by the different social subjects. Furthermore, it analyzes the dynamics of increasing territorial conflicts in the forested areas of the mountain region in both provinces, as a result of the advance of different public and private ventures on lands and territories occupied by the above-mentioned small producers.

Keywords Forests · Regulations · Territories · Policies

V. Iñigo Carrera (✉)
Universidad Nacional de Río Negro (UNRN) – Consejo Nacional de Investigaciones Científicas y Técnicas (CONICET), Instituto de Investigaciones en Diversidad Cultural y Procesos de Cambio (IIDyPCa), San Carlos de Bariloche, Argentina

A. Balazote
Universidad de Buenos Aires (UBA), Buenos Aires, Argentina

A. Balazote
Universidad Nacional de Luján (UNLu), Luján, Argentina

G. Stecher
Universidad Nacional del Comahue (UNComa), Asentamiento Universitario San Martín de Los Andes, Cátedra Extensión Rural, San Martín de los Andes, Argentina

1   Introduction

Since November 2007, the surface of native forests in Argentina has been managed by the State—at its different levels—pursuant to the established criteria for its enrichment, restoration, conservation, use and sustainable management, as well as those of the environmental services they provide to society, based on the provisions of National Law No. 26,331 on "Minimum Standards for the Environmental Protection of Native Forests" (known as the "Forest Law").[1] A proportion of this surface, corresponding to the northwestern region of the Andean-Patagonian forests, is located in the mountain area of the provinces of Neuquén and Río Negro (Argentine Northern Patagonia) (Fig. 1)[2] These lands are extremely rich, not only in terms of native forest mass but also soils, grasslands, waters, and landscapes. The traditional dwellers of these lands and territories are small producers who belong to —or who identify themselves as belonging to—indigenous peoples (in particular, the Mapuche people),[3] as well as those with a *criollo* origin.[4] National and foreign capitals have recently appropriated these lands and territories, particularly by means of consolidating the region for tourism and real estate expansion purposes (Blanco and Arias 2018; Iñigo Carrera 2019; Trpin and Rodríguez 2019; Valverde et al. 2015).

---

[1]Current legislation considers native forests as natural forest ecosystems mainly consisting of mature native tree species, with various species of associated flora and fauna, together with the environment that surrounds them (Art. 2, National Law No. 26,331). The tangible and intangible benefits generated by the native forest ecosystems are considered to be environmental services, namely water regulation; the conservation of biodiversity, soil and water quality; fixing of greenhouse gas emissions; the contribution to landscape diversification and beauty; the defense of cultural identity (Art. 5, National Law No. 26,331).

[2]Patagonia is located in the southern tip of Latin America, comprising the southern territories of Chile and Argentina. In Argentina, it covers part of the provinces of Buenos Aires, La Pampa and Mendoza, and the whole of the provinces of Neuquén, Río Negro, Chubut, Santa Cruz and Tierra del Fuego, Antarctica, and the South Atlantic Islands. This work focuses on the Northern Patagonian districts, mainly Neuquén and Río Negro. The Argentine Northern Patagonia region includes two types of subregions with differentiated geomorphological, climate, hydric, and vegetation features: the one located along the Andes mountain range, with the Andean-Patagonian forest on its slopes; and the other one with highlands and wide plateaus, where the Patagonian steppe is located.

[3]The Mapuche are one of the indigenous peoples that existed before the creation of the Argentine nation-state, a process based on the violent incorporation of those peoples. There are numerous academic of papers that conceptualize the expansive process of the national State and the policy applied to the indigenous peoples during the military conquest years (late nineteenth century) as a genocide (Delrio et al. 2010; Tamagno 2014; Trinchero 2006). A shift in the main production relations ensued, and the Mapuche inhabitants reproduced a precarious situation of land tenure which, to a great extent, remains unchanged until today, although they are increasingly organized in communities, *Lof* (collective subject based on family and land relations) and organizations based on claiming their specific ethnic identity and demanding the fulfillment of their acknowledged rights.

[4]*Criollos* are local, non-indigenous, residents of mainly European ancestry. They make up peasants communities.

Fig. 1 Native forests under the "Forest Law" in Neuquén and Río Negro, Argentina. *Source* Map made by the authors, based on data from the National Monitoring System of Native Forests of the Argentine Republic. Free software QGIS 3.12 Bucurestti. *Author* Gabriel Stecher

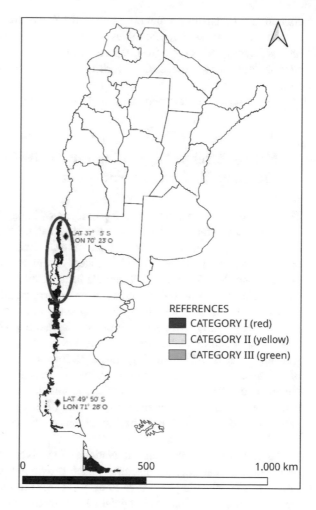

REFERENCES
CATEGORY I (red)
CATEGORY II (yellow)
CATEGORY III (green)

Our work examines the implications of public policies regarding nature conservation—in general—and territorial planning of native forests—in particular—in relation to the territorialities configured by the different social subjects. Furthermore, it analyzes the dynamics of increasing territorial conflicts in the forested areas of the mountain region of Northern Patagonia (in both provinces), as a result of the development of different public and private ventures on lands and territories occupied by small producers, on the one hand, and as a result of the lack of regulation of the producers' ownership of those lands and territories, on the other.

For this purpose, we have structured the chapter as follows. First, we present some general principles of the "Forest Law" and the respective laws by which the provinces of Neuquén and Río Negro abide. Next, we analyze the territorialities configured by the State at its different levels, capitals and small producers concerning the territorial planning policy for native forests and the territorial conflicts driven by these different logics, both in the province of Neuquén (Department of Los Lagos) and Río Negro (Department of Bariloche). Finally, we conclude by summarizing our approach. Our work is based on a literature review of empirical studies carried out in the area, analysis of secondary sources, participant-observation experiences, and interviews.

2 Native Forests: Legislation for Their Territorial Planning in Northern Patagonia

The surface of native forests in the provinces of Neuquén and Río Negro is managed pursuant to the provisions of National Law No. 26,331. This law was enacted in a context of intensive and extractive exploitation of natural resources, by emerging and multiplying conflicts related to this exploitation led by indigenous peoples and by different local populations (Giarracca 2006; Merlinsky 2013; Seoane 2006), and by the introduction of environmental considerations in public policies (Schmidt 2015). Unlike what happened in the Northern part of the country (in the Yungas area, Misiones rainforest, Espinal and Chaco parks), where deforestation assumed dramatic dimensions in the last two decades, Northern Patagonia was not subject to an equally high rate of deforestation resulting from the expansion of the agricultural frontier (mainly, soybean). Instead, the loss of native forests in Neuquén and Río Negro was one of the lowest in the country and was due to the occurrence of fires. Nevertheless, as the goal of protecting native forests grew to a national level, both provinces had to adhere to the terms set forth by the "Forest Law."

Pursuant to the provisions of National Law No. 26,331, within a maximum period of one year after its enactment, each provincial jurisdiction had to carry out the Territorial Planning of Native Forests (OTBN) in its own territory, following a participatory process and the established sustainability criteria, and configuring the conservation categories based on the environmental value of the different units of native forests and the environmental services provided by them.[5] The approval of each OTBN by law, and its certification by the National Secretariat of Environment and Sustainable Development (SAyDS), would allow the provinces to access the National Fund for the Enrichment and Conservation of Native Forests, as a compensation to the owners of the affected lands and to the provincial administrations for the conservation of the forests, acknowledging the environmental services they provide.

---

[5]Among the established sustainability criteria, it is worth mentioning criterion No. 10: Value given by the indigenous communities to forested areas and their surroundings, and the use of their natural resources for the purpose of their survival and maintaining their culture. In the "Forest Law," there are various references to the rights vindicated by the indigenous peoples who live or carry out activities in forested areas (Valtriani and Stecher 2019).

Thus, in November 2011, Neuquén passed its Provincial Law No. 2,780 on "Territorial Planning of Native Forests." In July 2010, Río Negro followed suit, passing Provincial Law No. 4,552 on "Conservation and Sustainable Use of Native Forests." In both jurisdictions, the largest amount of native forest surface was included in Category II (yellow—medium conservation value), a lower percentage in Category I (red—high conservation value), and a minimum proportion of that surface was included in Category III (green—low conservation value) (Table 1).[6]

3 Neuquén: Territorial Planning of Native Forests

In Neuquén, Provincial Law No. 2,780, passed in 2011, and its subsequent OTBN map, which covers an area of approximately 543,917 ha (Figs. 2 and 3), have sparked much debate since their enactment and provincial enforcement, which included the participation of environmental and social organizations, unions and small producers (indigenous and peasant communities). Following this process, the level of conflict and dispute between the different social subjects involved has increased, as reported by local, regional, and national media (radio, television, and printed or online newspapers).

The State's need to respond to the interests and pressure of dominant groups linked to real estate investments has led to the enactment of regulatory decrees that go against the spirit of the law. Incidentally, they represent a breach of national, provincial, and municipal legislation—which often recovers international declarations and agreements—focused on indigenous and environmental matters, which results in the renewed exclusion and invisibility of the indigenous and peasant communities. Hence, they are denied territorial rights over forested areas of traditional use (Stecher 2013). This situation reproduces a regional development model based on tourism, with a strong extractivist imprint, that widens the social and environmental gaps (Encabo et al. 2016) and considers the landscape and its territorial multi-dimensionalities as a commodity.

Although the legislation provides for the revision and update of this OTBN every five years, the local Enforcement Authority (the Forest Resources Division of the province of Neuquén) has modified the conservation categories, under the figure of "adjustments," even to the extent of excluding areas which were contemplated in the original maps. As we have previously mentioned, these changes are mainly based on real estate development projects, generating strong controversies among the different social subjects. In 2018, the Enforcement Authority opened the participation and consultation instances, as established by law, by holding workshops in the towns located in the mountain area, thus gathering citizens, socio-environmental organizations, and the scientific-technological sector. The

---

[6]It should be noted that they are among the provinces with the lowest area of native forests declared.

Table 1 Total surface of native forest, divided by conservation category, according to the OTBNs of the provinces of Neuquén and Río Negro, in Argentina

Province	Total surface (Ha)	Surface category I		Surface category II		Surface category III	
		Ha	%	Ha	%	Ha	%
Neuquén	543,917	192,686	35	347,672	64	3,559	1
Río Negro	478,900	181,900	38	252,700	53	44,300	9

Source Ministry of Environment and Sustainable Development of Argentina (MAyDS) (2017). *Author* Gabriel Stecher
Ha: hectares

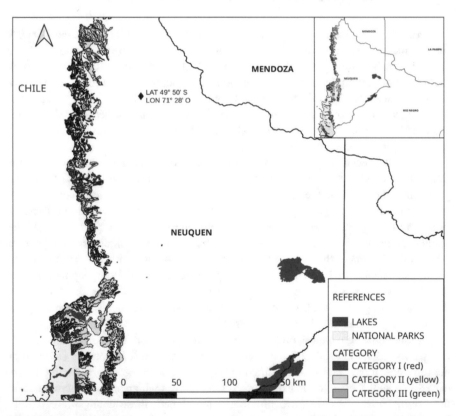

Fig. 2 Territorial Planning of Native Forests (OTBN), province of Neuquén (Northern area), Argentina. *Source* Map made by the authors, based on data from the National Monitoring System of Native Forests of the Argentine Republic. Free software QGIS 3.12 Bucurestti

methodological proposal was to focus on the representation and construction of new maps, making changes in the conservation categories based on the participants' perspectives, and thus creating spaces to visualize and reveal the different perceptions on which society—both concretely and symbolically—builds its decisions

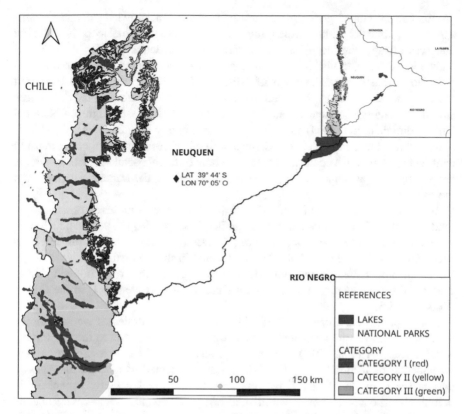

Fig. 3 Territorial Planning of Native Forests (OTBN), province of Neuquén (Southern area), Argentina. *Source* Map made by the authors, based on data from the National Monitoring System of Native Forests of the Argentine Republic. Free software QGIS 3.12 Bucurestti

about native forests as a common good (Arach and Stecher 2019). Due to the social heterogeneity of those spaces, different positions and analyses emerged—with various levels of agreement—that explicitly revealed the tension about forest conservation in the face of hegemonic development models.

A clear example of what has been described so far, also representative of other situations occurring at the provincial level—as Neuquén is characterized by an important social fabric made of indigenous and peasant communities—is the case of *Lof Paichil Antriao*. This Mapuche community is located on the slopes of the Belvedere Hill, within the 2-km municipal land surrounding the center of Villa La Angostura (Department of Los Lagos). Over the years, the size of its territory has been considerably reduced, since much of the old Plot 9 (ancestral location) is currently a luxurious residential neighborhood that encompasses important buildings used as households and for tourism purposes.

Since 2003, *Lof Paichil Antriao* has been conducting a series of mobilizations and judicial claims in search of recognition of its status as part of an indigenous group. The provincial government, however, has not acknowledged these claims and much public debate has ensued, regarding indigenous pre-existence in a locality that was always considered as not having any indigenous peoples (García and Valverde 2007). The visibility of these groups has prompted a territorial reconfiguration, transforming the town of Villa La Angostura itself into a contested territory and, as quoted by the aforementioned authors, exacerbating the differences among various government levels, since the Federal State—through the National Parks Administration (APN) and the National Institute of Indigenous Affairs (INAI) —has acknowledged the Mapuche demands. In other words, a process of increasing conflict has started in a contested territory, where different social subjects converge, crossed by deep asymmetries of power (private capitals, the Mapuche community, the State).

In 2017, the Enforcement Authority of the Forest Law authorized opening up a road—of about 1300 m long, at an average height surpassing the 900 m benchmark —to have access to a real estate venture ("Correntoso Lake Urbanization and New Waterfront" or "Correntoso Lake Trust") located in the area called Pichunco (Plot 9), in the upper area of the Belvedere Hill. Opening this road implied felling 1.34 ha of forests within a maximum conservation area (Category I) in the OTBN (Figs. 4 and 5).

For its part, if the urbanization of 87 ha surface were fulfilled, it would alter the Correntoso Lake basin, a territory that has traditionally belonged to the *Lof Paichil Antriao* (Stecher et al. 2018). Although the answer has been to establish an "environmental protection" area,[7] this action was prosecuted as a crime of unlawful appropriation by the State. However, part of the local society, organized in neighborhood councils, non-governmental organizations, and socio-environmental assemblies, started not only defending the community but also getting involved in the rights guaranteed in the Forest Law (Fig. 6).

On the other hand, the actions based on interpretative intentions, carried out by the Enforcement Authority in favor of real estate stakeholders' groups, created a context of greater social, cultural, and environmental conflict. This not only caused damage to the *Lof Paichil Antriao* but also created a perception of impunity in other local social groups. Thus, in a clear expression of joint and solidary action within the framework of the participatory instances established by National Law 26,331 to update the OTBN, the social groups managed to keep and even increase the protection categories, against a more lenient new proposal presented by the Enforcement Authority.

---

[7]It should be noted that private plotting will have a serious impact on biodiversity—called *ixofijmogen* in Mapuche language (*Mapuzugun*), which translates as "all the lives." Therefore, and considering that "the Mapuche people depends exclusively on it, and we could not develop ourselves as a culture without these various elements that biodiversity offers us, such as lakes, waterfalls, streams, rivers, swamps, menukos (water sources), medicinal plants, native tree species, stones, rocks, land, fauna, biotic species, etc." (Stecher et al. 2018), the plotting will have an equally serious impact on the social life of *Lof Paichil Antriao*.

Fig. 4 Deforestation in the area called Pichunco, Villa La Angostura, province of Neuquén, Argentina, 2018. *Source* Photograph taken by Gabriel Stecher

Fig. 5 Opening up a road in the area called Pichunco, Villa La Angostura, province of Neuquén, Argentina, 2019. *Source* Photograph taken by Gabriel Stecher

Fig. 6 Entrance to the "environmental protection" area, Villa La Angostura, province of Neuquén, Argentina. *Source* Photograph taken by Gabriel Stecher

4 Río Negro: Conservation and Sustainable Use of Native Forests

As it was previously mentioned, there are 478,900 ha of native forests managed pursuant to the provisions of Río Negro Provincial Law No. 4,552, through which Río Negro province abided by the terms established by the Forest Law. It addresses a surface located in the mountain strip to the Southwest of the province (Department of San Carlos de Bariloche). We also mentioned that according to the first OTBN conducted in the jurisdiction in 2010, 38% of the native forest area was included in Category I (red), 53% in Category II (yellow) and the remaining 9% in Category III (green) (Fig. 7).

This first zoning process—developed only by the province—was reviewed in February 2015, i.e., long after the two years established by provincial law to conduct the review had gone by, and even today such review has not been submitted for approval to the provincial Parliament. The review process, which started in February 2011, was conducted by the Enforcement Authority, the Provincial Enforcement Unit for the Protection of Native Forests (UEP-PBN), through the Advisory Council in the Andean Area, created by the Unit.[8] The role of this Council was to advise the Forest Division of the provincial Ministry of Agriculture, Livestock and Fisheries (MAGyP) and the provincial Secretariat of Environment

---

[8]For a detailed analysis of the Advisory Council's actions on the social management of the territory, and in particular, an analysis of the actors (their interests, positions, demands) involved in the participatory process led by this consultative-participatory body, see Namiot (2018).

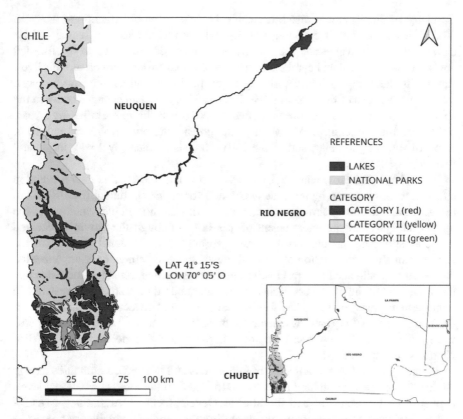

Fig. 7 Territorial Planning of Native Forests (OTBN), province of Río Negro, Argentina. *Source* Map made by the authors, based on data from the National Monitoring System of Native Forests of the Argentine Republic. Free software QGIS 3.12 Bucurestti

and Sustainable Development (SAyDS) on matters related to the effective implementation of Provincial Law 4,552 in the territory, although its judgment was not binding. The Council consists of public and private institutions interested in the conservation of native forests, and it focused on seven core areas: governmental; academic and research; intermediate non-governmental organizations; professional associations; indigenous peoples; primary producers and rural dwellers; forest service providers and related industries. Among the indigenous peoples, different Mapuche communities participated in that instance: *Huaytekas, Tacul, Huenchupan, Huenuleo, Quijada, Ranquehue, Tripay Antu, Follil, Rayen Mapu, Ñirihuau* (Namiot 2018).

The rezoning process under the OTBN was based on the revision of the criteria used in 2010 with proposals to modify or to include a new one. In this sense, the first OTBN did not consider all the sustainability criteria set forth in the law: the one related to biological aspects had a greater significance, while those related to the value indigenous communities attach to native forests was not taken into account

(Ministry of Environment and Sustainable Development of Argentina—MAyDS 2017). However, the 2015 zoning process did not include some essential criteria either, such as the value and the use given by these communities to the forested areas for the (concrete and symbolic) purpose of social (re)production of their lives. Rather, its definition mainly focused again on the biological aspects and, at most, the productive uses of those areas (UEP-PBN 2013). The main changes between the territorial zoning of native forests drafted in 2010 and the one drafted five years later were the reduction of the yellow and green categories (by 32% and 16%, respectively), in particular, and of the native forests surface (by 13%), in general (Namiot 2018).

The delay in the provincial Legislature's approval of the 2015 review resulted in a formal warning issued by the Federal Council for the Environment (COFEMA) to the governor. It also occurred simultaneously with the non-implementation of the budget allocated for the enforcement of this law. Although the province received the corresponding annual compensation through the National Fund for the Enrichment and Conservation of Native Forests, the government only implemented the percentage allocated to forest institutional strengthening, retaining the funds that had to be sent to the producers. Hence, over one hundred plans (mainly, sustainable management plans) presented by forested land owners between 2010 and 2015 for the harnessing of the forest resources (both its timber and non-timber resources) were left without the necessary approval and financing to be implemented (Río Negro Newspaper 2018).

In line with the low participation of indigenous and peasant communities among the landowners who submitted plans for 2015 (1.3% of the total, at a national level) (Ministry of Environment and Sustainable Development of Argentina—MAyDS 2017), few communities—among the seven involved in the participatory process— submitted a management plan.[9] One of them was *Las Huaytekas*, settled in the rural areas of El Foyel and Los Repollos (about 70 km from the city of San Carlos de Bariloche and 40 km from the town of El Bolsón, in the Department of Bariloche). Its territory, rich in forest species, has led both to the search for its conservation— through National Law No. 26,331 as well as through previous legislation—and to its exploitation by capitals focused on industrial logging.

Therefore, on the one hand, this community's territory is now almost entirely included within the limits of *Cipresal de las Guaitecas* Protected Natural Area (ANPCG). The area (around 4,000 ha) was created by Provincial Law No. 4,047 in 2005, to ensure the conservation of the relict species of Guaitecas cypress, mountain cypress, and larch (all of them, endangered or vulnerable species) and of the biological diversity and the ecological and natural evolutionary processes that create and/or affect this relict. The cypress area hosts activities related to the

---

[9]Technical difficulties faced when completing the required administrative procedures might explain the low proportion of plans implemented by indigenous communities. The "Forest Law" foresees that a part of the 30% of the fund granted to the provinces is used for technical and financial assistance to small producers (indigenous and peasants communities). However, it is unknown to what extent this was actually implemented (Aguiar et al. 2018).

material subsistence of the community (obtaining water, grazing animals, collecting mushrooms and medicinal plants). It also includes their *rewe* (spiritual and cere-monial site), where the *kamaruko* (a Mapuche ceremony) is performed. Different judicial decisions have recognized the significance of the cypress area for the reproduction of their traditional practices and identity as a people, and the com-munity's rights over that space (Amico et al. 2014). Although the Protected Natural Area was created without prior consultation or participation and, even more so, with opposition from the community (Las Huaytekas 2009), an agreement was reached later on with the Council of Ecology and Environment (CODEMA) to implement a management plan for the cypress area. Eight years after the start of its elaboration, this plan has not been approved yet.

On the other hand, private and State capitals claimed to be interested in implementing afforestation plans, to justify the acquisition of fractions of fiscal lands—often through irregular transfers—that overlapped with the community's traditional territory. Incidentally, forestation initiatives with fast-growing exotic conifers (whether it implied clearing the native forests or not) have been promoted by the national and provincial States since the 1970s and led by private and mixed capitals (with State participation). This activity has not been highly developed: the forested area in Río Negro amounts to barely 11% of the forested area in the northwestern region of Patagonia (around 11,860 ha), a much lower figure than in Neuquén province (National Service of Agri-Food Health and Quality—SENASA 2017). However, afforestation has been pointed out by Mapuche communities and organizations as a tool used by those capitals for the appropriation of their territory (Consejo Asesor Indígena—CAI 2011).[10] Thus, on the one hand, the indigenous groups that organize their production based on subsistence work find it difficult to access the necessary transportation guides for the use and/or commercialization of forest products (firewood and wood), since their occupied and/or claimed territories are not effectively recognized. On the other hand, companies focused on wood production extract these resources for their use and benefit, or they change their initial purpose of wood production and become tourism and real estate enterprises,

---

[10]In this sense, the afforestation processes carried out by the Forestry Company of Río Negro (EMFORSA) (of mixed capitals, and focused on the afforestation, industrialization, and com-mercialization of forest raw materials) are paradigmatic. These capitals have been involved not only in land transfers to private hands, but also in an irregular transfer. In 1999, around 289 ha of fiscal land were transferred to EMFORSA. Then, the State contributed its own land as capital. One year later, when the company was in a financial deficit situation, that parcel of land was sold—at its assessed value—to the former chairperson of the company between 1998 and 2012, who owned an adjacent parcel of land. He gained ownership only in 2001, after starting an eviction trial against an occupier of the fiscal land. That same year, he transferred the lands to Hidden Lake (a British corporation that owns 11,284 ha in the Department of Bariloche) (Iñigo Carrera 2020). Such land is part of the *Lof* Palma-Villablanca territory, a member of *Las Huaytekas* community.

within the framework of an ongoing increase of land prices since the late 1980s—
with even greater intensity during the 2000s—particularly in tourist fractions of
land, on the river or lake shore.[11]

In this sense, the defense of the occupied and/or claimed territory, and the access
and control of its resources—including, among others, the possibility of firewood
extraction, a typical activity among indigenous and non-indigenous small producers
—are the cornerstones on which *Las Huaytekas* community has based its main
political actions for the last eight years (land recoveries, installing a gate to prevent
entry to the territory, festivals, peaceful occupation of provincially-managed
buildings, celebration of *kamarucos*, legal appeals). These actions have gained
visibility on both the provincial and the national levels.[12]

5 Final Remarks

Throughout our work we presented general principles set forth by National Law
No. 26,331 on "Minimum Standards for the Environmental Protection of Native
Forests" (known as the "Forest Law") and the respective laws that Neuquén and Río
Negro provinces abided by. We then analyzed the territorialities configured by the
State, the capitals and small producers in relation to the territorial planning of native
forests—in particular—and the conservation of nature—in general—as well as the
territorial conflicts sparked by these different logics, both in the departments of Los
Lagos (Neuquén) and Bariloche (Río Negro).

These conflicts stem from the development of different public and private
ventures on the lands and territories occupied by the small producers who belong to
—or identify themselves as belonging to—the indigenous peoples, as well as those
with *criollo* origin. Thus, public policy regarding the definition of territorial planning of native forests is pressed between the protection and conservation of natural
resources, on the one hand, and the valuation of industrial capitals applied to
different productive activities, including tourism and forestry, on the other. It is a
tension that is contained in it but also transcends it, and that emerges in the form of
the above-mentioned conflicts. In other words, in analyzing the implications of this
public policy in relation to the territorialities configured by the different social

---

[11]One of the reasons given by the then Deputy Secretary of Forest Resources of the Ministry of
Agriculture, Livestock and Fisheries of the province—when submitting his resignation in 2014—
for the delay in submitting the new OTBN, established for 2012, was the dispute about the zoning
classification as Category III (green) or Category II (yellow) of the land parcels included in the
Cerro Perito Moreno Comprehensive Development Project (Agencia Digital de Noticias 2014).
This project is a tourism and real estate development one, around 25 km away from El Bolsón,
located within different nature conservation areas, which entailed the development of a ski center
and a tourist village at the base of the hill (Iñigo Carrera 2019).

[12]This is common to other communities throughout Patagonia (Briones and Ramos 2020;
Schiaffini 2019; Tozzini 2014; among others).

subjects, we must reconsider the question about the relationship between nature conservation and the (re)creation of the necessary conditions for capital accumulation, through the dispossession of land and natural resources, the exclusion of local populations, and the promotion of nature-based tourism, among other mechanisms. Although at first sight it may seem contradictory, the link between these aspects is not necessarily obvious or immediately perceived.

References

Agencia Digital de Noticias (2014, September 1st) Renunció Javier Grosfeld a la Subsecretaría de Recursos Forestales. Last access: 9 Oct 9 2020. https://adnrionegro.com.ar/2014/09/renuncio-javier-grosfeld-a-la-subsecretaria-de-recursos-forestales/

Aguiar S, Mastrangelo M, García Collazo M, Camba Sans G, Mosso C, Ciuffoli L, Schmidt M, Vallejos M, Langbehn L, Cáceres D, Merlinsky G, Paruelo J, Seghezzo L, Staiano L, Texeira M, Volante J, Verón S (2018) ¿Cuál es la situación de la Ley de Bosques en la Región Chaqueña a diez años de su sanción? Revisar su pasado para discutir su futuro. Ecología Austral 28:400–417

Amico G, Briones C, López Alaniz P, Nahuelquir F, Ñancunao M, Paritsis J, Puntieri J, Ruiz C (2014) Informe final. Elaboración e implementación de un plan de manejo preliminar del Área Natural Protegida Cipresal de las Guaitecas (ANPCG). San Carlos de Bariloche

Arach A, Stecher G (2019) Proceso de revisión del OTBN en el sur de Neuquén. Experiencias de participación de los actores socioterritoriales. IV Jornadas Forestales de Patagonia Sur. IV Congreso Internacional Agroforestal Patagónico. Universidad Nacional de Tierra del Fuego, Ushuaia

Blanco G, Arias F (2018) Las comunidades indígenas en Neuquén y la política de tierras en el largo plazo (fines del siglo XIX – fines del siglo XX). In: Blanco G (ed) La tierra pública en la Patagonia. Normas, usos, actores sociales y tramas relacionales. Prohistoria, Rosario, pp 227–257

Briones C, Ramos A (2020) Los porqués del "de acá nos van a sacar muertos". Procesos de recuperación de tierras en la Patagonia Norte. Intersticios de la política y la cultura. Intervenciones latinoamericanas 9(17):9–43

Consejo Asesor Indígena (CAI) (2011, May 6th) La forestación como instrumento de apropiación del territorio en Cuesta del Ternero. Last access: 15 Oct 2020. http://argentina.indymedia.org/news/2011/05/778965.php

Delrio W, Lenton D, Musante M, Nagy M, Papazian A, Pérez P (2010) Discussing indigenous genocide in Argentine: past, present, and consequences of Argentinean state policies toward native peoples. Genocide Stud Prev Int J 5(2):138–159

Encabo M, Sánchez S, Torre G, Paz Barreto D, Andrés J, Mastrocola Y, Vázquez V, Cánepa L (2016) Uso responsable de biodiversidad. Revisando el modelo recreación y turismo en conservación. Anuario de Estudios en Turismo. Investigación y Extensión 16(XI):8–20

García A, Valverde S (2007) Políticas estatales y procesos de etnogénesis en el caso de poblaciones mapuche de Villa La Angostura, provincia de Neuquén, Argentina. Cuadernos De Antropología Social 25:111–132

Giarracca N (2006) Territorios en disputa: los bienes naturales en el centro de la escena. Realidad Económica 217:51–68

Iñigo Carrera V (2019) Relaciones capitalistas y conflictos territoriales: una aproximación a su emergencia y desarrollo en la cordillera rionegrina. In: Kropff L, Pérez P, Cañuqueo L, Wallace J (eds) La tierra de los otros. La dimensión territorial del genocidio indígena en Río Negro y sus efectos en el presente. Universidad Nacional de Río Negro, Viedma, pp 185–215

Iñigo Carrera V (2020) Las formas del despojo en la cordillera rionegrina: a propósito de las trayectorias de dos empresas forestales, Pilquen. Sección Ciencias Sociales 23(2):14–28. Last access: 9 Oct 2020. http://revele.uncoma.edu.ar/htdoc/revele/index.php/Sociales/article/view/2630

Las Huaytekas (2009, September 23rd) Comunicado para la prensa. Last access: 9 Oct 2020. https://www.barilochense.com/bariloche-social/pueblomapuche/rio-negro-comunidad-mapuche-las-huaytekas-recupera-territorio

Merlinsky G (2013) Introducción. La cuestión ambiental en la agenda pública. In: Merlinsky G (Compilator) Cartografías del conflicto ambiental en Argentina. CICCUS–CLACSO, Buenos Aires, pp 19–60

Ministerio de Ambiente y Desarrollo Sustentable de la Nación (MAyDS) (2017) Ley N° 26.331 de Presupuestos Mínimos de Protección Ambiental de los Bosques Nativos. Informe de estado de implementación 2010–2016. Last access: 9 Oct 2020. https://www.google.com.ar/url%3Fsa%3Dt%26rct%3Dj%26q%3D%26esrc%3Ds%26source%3Dweb%26cd%3D1%26ved%3D2ahUKEwj6v4zksdHnAhV_E7kGHTWyB-cQFjAAegQIAxAB%26url%3D https://www.argentina.gob.ar/sites/default/files/informe_de_implementacion_2010_-_2016.pdf%26usg%3DAOvVaw1Ojs4KpTebNhfGGltDup9G

Namiot G (2018) Percepciones, intereses y relación de fuerzas en el Ordenamiento Territorial de los Bosques Nativos. Estudio de caso: Consejo Consultivo en El Bolsón, Provincia de Río Negro. Unpublished master thesis, Universidad Nacional del Comahue

Río Negro Newspaper (2018, August 5th) Río Negro incumple la ley de Bosque Nativo y retiene fondos. Last access: 9 Oct 2020. https://www.rionegro.com.ar/rio-negro-incumple-la-ley-de-bosque-nativo-y-retiene-fondos-BD5511355/

Schiaffini H (2019) Conflictividad rural, estructura social y etnicidad en Chubut. Las "recuperaciones territoriales mapuche" en perspectiva social e histórica. Entramados y perspectivas 9(9):3–32

Schmidt M (2015) Política ambiental, avance de la frontera agropecuaria y deforestación en Argentina: el caso de la ley "De Bosques." GeoPantanal 18:121–139

Seoane J (2006) Movimientos sociales y recursos naturales en América Latina: resistencias al neoliberalismo, configuración de alternativas. Sociedade e Estado 21(1):85–107

Servicio Nacional de Sanidad y Calidad Agroalimentaria (SENASA) (2017) Anuario estadístico 2016. General Roca. Last access: 9 Oct 2020. http://www.senasa.gob.ar/institucional/centros-regionales/centros-regionales/patagonia-norte

Stecher G (2013) Ley de Bosques. Su aplicación en territorios de comunidades campesinas e indígenas en la Provincia de Neuquén. Nuevos modos de exclusión. VII Jornadas Santiago Wallace de Investigación en Antropología Social. Universidad de Buenos Aires, Buenos Aires

Stecher G, Arach A, Nahuel F, Lonkon L (2018) Informe apertura camino Lote 9 sector Pichunco. Universidad de Buenos Aires, Buenos Aires

Tamagno L (2014) Políticas indigenistas hoy. Un nuevo «parto de la antropología». Etnicidad y clase. In: Trinchero H, Campos Muñoz L, Valverde S (Coordinators) Pueblos indígenas, conformación de los estados nacionales y fronteras. Tensiones y paradojas de los procesos de transición contemporáneos en América Latina. Editorial de la Facultad de Filosofía y Letras, Buenos Aires, pp 9–35

Tozzini A (2014) Pudiendo ser mapuche. Reclamos territoriales, procesos identitarios y Estado en Lago Puelo, provincia de Chubut. Universidad Nacional de Río Negro, Viedma

Trinchero HH (2006) The genocide of indigenous people in the formation of the Argentine nation-state. J Genocide Res 8(2):121–135

Trpin V, Rodríguez MD (2019) Transformaciones territoriales y desigualdades en el norte de la Patagonia: extractivismo y conflictos en áreas agrarias y turísticas. Albuquerque. Revista de Historia 10(20):50–66

Unidad Ejecutora Provincial de Protección de Bosques Nativos (UEP-PBN) (2013) Criterios de zonificación de Ley de Protección de Bosques Nativos 4.552. Consejo Consultivo de Bosques Nativos. Resumen de acuerdos y desacuerdos. San Carlos de Bariloche

Valtriani A, Stecher G (2019) Aspectos cualitativos y cuantitativos. Comparativos de su aplicación en dos provincias patagónicas. IV Jornadas Forestales de Patagonia Sur – IV Congreso Internacional Agroforestal Patagónico. Universidad Nacional de Tierra del Fuego, Ushuaia

Valverde S, Maragliano G, Impemba M (2015) Expansionismo turístico, poblaciones indígenas Mapuche y territorios en conflicto en Neuquén, Argentina. Pasos. Revista de Turismo y Patrimonio Cultural 13(2):395–410

Indigenous and Peasants Lands Under the Spotlight—A State Forest Policy in the Gran Chaco, Argentina

Natalia Castelnuovo Biraben

Abstract This chapter analyzes national forestry policy focused on peasants and indigenous peoples who inhabit the Gran Chaco, Argentina, whose lands and life systems have been affected by deforestation, the expansion of the livestock farming model, and what some authors call "world land fever." The forestry policy under analysis is the Native and Community Forestry Project (2015–2020), managed by the Undersecretariat of Planning and the Environment and Sustainable Development Secretariat in close collaboration with the Forestry Department. The main purpose of this work is to highlight how and in which ways the "strengthening of land tenure" is taken as a pivotal objective of the forestry policy, which arc some of the procedures (accreditation, minutes, administrative records, guarantees, etc.), and State regulation methods brought into play from the perspective of officers and project technicians. In the framework of "strengthening," the project implemented an organization of the legal land registry of the territories together with a series of State provincial agencies. It is important to point out the contradictions and tensions that arose during the policy implementation, in relation to the importance acquired by some of the procedures and requirements deployed by state agencies in order to endorse the project [Research for this chapter took place within the UBACyT Research Project (20020170100457BA) "Practical articulations of different political-administrative levels of organization: social relationships and political processes" (2018–2021), and within the framework of the FILOCyT Program (FC19-015) "Public policy, territorialities and technological devices: a comparative anthropological analysis of social processes of production, demarcation and spatial representation in Argentina" (2018–2021)].

Keywords Forests · Public policies · Public agencies

N. Castelnuovo Biraben (✉)
Consejo Nacional de Investigaciones Científicas y Técnicas (CONICET), Facultad de Filosofía y Letras, Universidad de Buenos Aires, Buenos Aires, Argentina

1   Introduction

The Native Forests and Community Project (from here on, Forests project) is a state forest policy directed to peasants and indigenous peoples of the Argentine Gran Chaco,[1] whose lands and life systems are being affected by deforestation, the expansion of agribusiness and what certain analysts call the "global land rush" (Cotula 2012). The project focuses on areas in the provinces of Salta, Santiago del Estero and Chaco,[2] and particularly on specific departments that have been affected the most by the expansion of agriculture and livestock farming and the dramatic increase of deforestation over the last decade.[3] However, these areas are also the ones with the highest density of native forests in the country and are home to thousands of peasant families and indigenous communities. These characteristics were essential when defining targeted areas, and they became part of the "criteria" to define policy recipients.

---

[1]The Argentine Gran Chaco comprises the provinces of Formosa, Chaco, Santiago del Estero, north of Córdoba, Santa Fe, San Luis, and east of Salta, Tucumán, Catamarca, La Rioja, San Juan and northwest of Corrientes. The Gran Chaco region geographically limits from north to south with the San Jose and San Carlos mountain chain at the southeast of Bolivia, up to the Salado river in Argentina, and from west to east with the last Argentine-Bolivian Subandean foothills up to the Paraná and Paraguay rivers. It is a vast, mostly semiarid, flatland of about a million square kilometers that comprises parts of Bolivia, Paraguay, Brazil and Argentina. The European conquerors who arrived in the sixteenth century "encountered" many indigenous hunters, gatherers, fishers, and cultivators that practiced seasonal horticulture, who inhabited this region and continue to do so. Among others, the Qom (Toba), the Nivacklé (Chulupí), Chané, Guaraní, Iyowaja (Chorote), Tapi'y (Tapiete) and Wichí people.

[2]These provinces are located in the north of the country. Historically, forest exploitation has been the main economic activity in Santiago del Estero. In 2017, the extraction of native forest products amounted to 600 thousand tons, representing 15% of the national total. The production of carbon is the province's main forestry activity. Santiago is the second carbon-producing province after Chaco. Agriculture and livestock farming have also expanded and play a significant role in the province, mainly between the Northern Salado and Dulce rivers (Provincial Production Report, Santiago del Estero 2019). Salta's production matrix presents a structure focused on the primary sector: agriculture and livestock farming and mining activities, and an important hydrocarbon industry. The province is also a significant tourist attraction (Productive Provincial Reports, Salta 2017). The province of Chaco has been the country's largest cotton producer. However, in the last years, it has fallen second to Santiago del Estero, due to the expansion of maize, sunflower and soybean. In fact, soybean is now the province's main crop. Forestry in Chaco is based on the exploitation of native species to provide for the tannin and railway tie industries. Over the last two decades, the province's cattle stock has increased. Chaco and Santiago del Estero have become the provinces with the highest cultivated areas with grains and oleaginous crops outside of the Pampean area (Provincial Production Report, Chaco 2019).

[3]Several authors have described and reported about the expansion of the agricultural border in certain provinces of the Argentine north and its social-environmental effects (See, for example, Aguiar et al. 2018; Camba Sans et al. 2018; Grau et al. 2005; Hansen et al. 2013; Nolte et al. 2017; Torella et al. 2018; Venencia et al. 2012).

In the Gran Chaco the precarious forms of land tenure,[4] the weak protection of rights[5] and large-scale agricultural investments enable capital concentration and land grabbing processes. Consequently, indigenous and peasant communities and/ or small producers are increasingly evicted from their lands, as territorial disputes increase. Around 95% of the forest lands of the *Parque Chaqueño* are involved in conflicts over tenure rights (Venencia et al. 2012) and 60% of the region's indigenous communities lack property titles of their ancestral lands (Observatorio de Políticas Públicas 2012).

There was yet another reason why the Forests project focused on these actors as recipients. After more than six years since the sanction of National Law 26,331 on Minimum Standards for the Environmental Protection of Native Forests, and five years since its regulation, only very few indigenous and peasant communities had been able to access to the National Fund for the Enrichment and Conservation of Native Forests.[6] Through this Fund, indigenous and peasant communities were expected to create their own Native Forests Management and Conservation Plan.

The Fund aims to "compensate jurisdictions that preserve native forests for the environmental services they provide"[7] (National Law 26,331, section 11). According to the provincial authorities in charge of the Fund's implementation, indigenous and peasant communities had not been prioritized as beneficiaries due to "the requirement of land tenure" (Audit Report AGN 2017: 40), a requirement that

---

[4]Di Paola and Ramírez (2014) classified precarious forms of land tenure as follows: untitled *criollos* (possessors or holders) in public or private-owned land; untitled *criollos* with 20 years of possession of public or private-owned land; untitled indigenous community on public or private-owned land; title-holding indigenous community and overlapping with families on public, private-owned or mixed land and indigenous community with insufficient lands on public land.

[5]The weak protection and implementation of indigenous people's rights contrasts with the vast body of current regulations in which the Argentine state acknowledges their right to lands that are sufficient and adequate for their development. The Civil Code, in several of its articles (2.351, 3.948, and 4.015), recognizes the right of inhabitants to land ownership when they have exercised peaceful, continuous, and uninterrupted possession for over twenty years, making improvements and investments, and when they have acted "with the intention of an owner." Despite the Code's legislation, peasants have been subject to evictions over the last fifteen years (De Dios 2012). In response to this problem, and to curb foreign ownership of rural lands, national and provincial-level regulations and agencies were created to conduct a rural lands register, to regularize land tenure, suspend evictions of small peasant and indigenous producers and intervene in conflicts over land issues.

[6]They receive technical assistance for planning activities concerning the protection, restoration, and use of their forest resources within a land-use planning model by which they identify preservation, enclosed and harvest areas, and areas for agriculture and livestock farming.

[7]Compensation refers to economic support. The benefit is a non-reimbursable contribution, paid per hectare and per year, according to the forest categorization.

the vast majority of indigenous and peasant communities cannot comply with. The provincial jurisdictions should allocate 70% of the budget they receive through the Fund to compensate land title-holders[8] (public or private), for their preservation of native forests, according to the conservation categories (Art. 35, Forest Law). The National Forests Administrator referred to this situation during an interview in 2019:

> 70% of the Fund's budget allocation to each province goes to beneficiaries. The beneficiary is an inhabitant, a citizen that owns forests. In many cases these forests are in private properties. Many of them are also located on public land [...] In the north there are mostly non-titled lands or irregular land tenure systems, and this is where we have the largest concentration of native forests. It is [also] where communities most use the native forest as a resource and a way of life, in contrast to other less densely populated regions.

Thus, the main goals of the forestry policy were outlined based on the convergence of all these factors, according to members of the executing unit of the Forests project at the national level. The different actors and agencies at stake—the World Bank, the national and provincial states, NGOs, and small peasant and indigenous producers—weighed and valued these goals differentially according to their diverse interests and expectations.

According to the National Secretariat of Environment and Sustainable Development and the Forests Administration, the Forests project's main goal was to "strengthen the Law" by improving community participation through Integrated Community Plans (*Planes Integrales Comunitarios*, from now on PIC, for its acronym in Spanish), and to transfer lessons learned and accomplishments to the Forest Law. In contrast, the team of formulators and technicians of the Forests project's national executing unit viewed the project's main goal as "the strengthening of land tenure through the Law" (interview with a national executing unit team member). The Integrated Community Framework, one of the project's key documents, partially states this goal. The document addresses the goal of strengthening land tenure and gives it new meaning. It stresses the importance of "strengthening the roots of the indigenous and *criollo* families that live in areas with native forests, improving their quality of life, and promoting the conservation, restoration, sustainable use and valuing of forest resources in some of the most environmentally and socially critical areas" (Integrated Community Framework, n. d. 54). This goal is linked to other broader ones that are mentioned in the Project Appraisal Document—which the formulators and technicians call the Project's Initial Assessment Document—and that refer to "improving forest management and increasing the access of small producers (including peasants and indigenous peoples) to markets and basic services in the selected northern provinces" (Project Appraisal Document 2015: 5).

---

[8]All people who have conducted their Native Forests Management and Conservation Plan and that have keep it updated—approved in each case by the application authority of the respective jurisdiction—may benefit from the Fund (Cabrol and Cáceres 2017). Landowners must have the Territorial Planning of Native Forests (OTBN) updated and approved by the provinces. The OTBN zones forests into conservation categories (red, yellow, and green).

According to the points of view of formulators and technicians of the project's executing team at the national level, the Forest policy's main aim was to strengthen land tenure. These actors perceived this strengthening as the chief way to consolidate small peasant and indigenous producers' land tenure.

The chapter analyzes State practices, procedures, and forms of regulation carried out during the project's implementation directed at organizing, regularizing, and acknowledging different forms of precarious land tenure. These actions entailed setting forth various mechanisms and practices: certifying the possession of community lands through guarantees, producing reports and administrative agreements, delimitating and demarcating land, creating maps, etc. Actions linked to strengthening land tenure were carried out by different provincial state agencies that oversaw putting into practice the "cadastral legal system" on certain project recipients' lands and territories. As a result, certifying the "non-conflictive nature of lands" and delineating the stretch of lands-territories in contexts that had never been previously surveyed by the State became essential project procedures and practices. Ethnographic analysis reveals the social actors' particular interest in the production and demarcation of territories, which evokes the notion of space as an unfinished historical product whose contradictions prevent that system from constituting itself and closing (Lefebvre 1974). Territories are configured through the interactions of different actors located in multiple and changing geographies, with divergent interests and expectations. The study reveals how the analyzed forest policy contributes to the Nation-State's objective of regularizing, organizing and imposing production practices, uses and territorial delimitations. Understanding that the State's primary goal is to strengthen land tenure, one of the effects of its actions through the Forests policy is the production of new meanings and representations of space and territory. Rather than assuming a disjointed state and actions, this study offers empirical evidence on how the State, through the Native Forests project, imposes a set of bureaucratic procedures and practices instituted and naturalized as the only possible way of organizing and acknowledging the lands-territories of peasants and indigenous peoples.

Research for this chapter is based on a multi-sited ethnography (Marcus 1995) that explores the global connections created by NGOs between States and local populations, as well as the frictions (Tsing 2005). I conducted fieldwork in several locations: the city of Buenos Aires, the provinces of Santiago del Estero and Salta. In 2016, I interacted for the first time with actors participating in the Forests project, in the Rivadavia Department, in the province of Salta. At that time, I participated in workshops with indigenous communities that had been selected as project "recipients." Over time, my research also sought to include the perspectives of officials, formulators, and technicians in charge of the implementation of the forest policy. Within the executing team at the national level, I interviewed the project manager, the vice-manager, and the coordinator, as well as the person in charge of social and environmental safeguard. I interviewed different World Bank officials linked to the project, as well as officials from the National Forests Administration and consultants. In the provinces, I contacted and interviewed coordinators as well as State and NGO technicians that formulated the Integrated Community Plans. In addition to analyzing a significant amount of in-depth interviews, my research also included analyzing documents produced by the project itself: the Integrated Community Plans Manual, the Participation Guide for the Development of Integrated Community Plans, the Project Appraisal Document and the Integrated Community Framework. The analysis of these materials as part of a documentary collection was essential. I agree with Stoler's approach to archives as objects rather than as sources, that is, "not as sites of knowledge retrieval, but of knowledge production, as monuments of states as well as sites for state ethnography" (2010: 469). The value of a documentary collection, as affirmed by Das and Poole (2008), lies in that "so much of the modern state is constructed through its writing practices" (2008: 26). This study focuses, on the one hand, on the perspectives of formulators and technicians of the project's executing team at the national level and, on the other, on the actions of specific State agencies in the province of Santiago del Estero.

2   State Mechanisms of Strengthening Land Tenure: Scope and Restrictions

The Undersecretariat of Planning, the Forests Administration of this Undersecretariat, and the National Environment and Sustainable Development Secretariat are the State agencies in charge of managing the Forests project (2015–2020). Given the natural administration of the State, the project operated in rented offices at the local level giving various degrees of autonomy. The project was the result of a long preparation process that led to its approval in March 2015. It received funding from the World Bank through a 58.76-million-dollar loan from the International Bank for Reconstruction and Development (IBRD).

According to the national Forests Administration, one of the project's main goals is that PICs become a course of action, recognizing the logics of communal use of the territory. Also, underlying this goal is the idea that the PICs can be later incorporated as part of the activity and functioning of the Forest Law.[9] During an interview in 2019, the National Forest Administrator was eloquent in this respect:

> In relation to the Law's [N. 26,331] management, the government understands that the project's main goal is the design of this tool. [The aim] is to include this vulnerable sector [indigenous peoples and peasants], which the Law had originally considered, but eventually left out in its implementation. We want to offer an instrument – the Integrated Community Plan – so that the provinces and beneficiaries can apply to the Fund and access to it, and that is incorporated into the regular implementation of the Forest Law.

Formulators and technicians of the executing team at the national level value this goal. However, according to their view, the initiative's main potential was that it

---

[9]The Forest Law identifies a privately owned property plan defined, with a titleholder that uses it individually.

opened up the possibility of strengthening land tenure through management plans. In their way of seeing things, even though "the Law had not been envisioned with this goal," the plans could be ways of "setting a precedent,"[10] and they intended to give them this meaning. What is relevant here is that the national State sets a precedent regarding the occupation of "public lands" that undoubtedly are owned by the provincial State, and not by the national State. Formulators and technicians referred to this during an interview in 2018:

> To us, this is a space occupation project through administrative acts that set precedents. That being made clear, we can discuss whatever you want; we can discuss investments, apiculture investments, the commercialization of honey and whatnot. [But] my [main] concern is that people secure land tenure. This is what they lack, and it is what they can achieve through this project.

To the formulating team, it was noteworthy that the national government[11] did not place any obstacles or restrictions to moving forward with this line. They felt this had to do with the national government's lack of "understanding" of what was being done. In the words of a member of the formulating team:

> I think the national government [of the 2015-2019 period] never fully understood the nature of the project. It implied taking risks, working with the most challenging aspects of family agriculture... The project's main difficulty was the essential issue of land tenure. We are discussing a forests project here, but this is the case for any family agriculture project. And we faced this difficulty head-on (2019).

The Forests project exclusively targets indigenous and peasant communities and small producers, regardless of their tenure situation. Although a vast majority of inhabitants are in a situation of precarious land tenure (on occasions referred to as "imperfect titling") this was not an obstacle for "communities" to become project recipients. The World Bank accepted this condition from the formulating team's proposal. A Bank official considered it as...

> An infrequent feature in the project's formulation, or at least an uncommon one in a project funded by the Bank. The project's formulation explicitly stated that land titles were not going to be a requirement for beneficiaries to access some of the project's benefits.

From the beginning, the team of formulators at the national level understood that the project had to include the provincial governments in the management of forest initiatives to achieve its goals. Not only because, according to the existing legal institutional framework guaranteed by the National Constitution, provincial states

---

[10]Peasants and indigenous peoples use the physical landmarks that come with different initiatives (wire fences, infrastructure works, production improvements, etc.), as material evidence (proof) to account for forms of land occupation.

[11]Preparation for the project began in 2013, during the administration of Cristina Kirchner (2007–2015). Shortly after the project's approval, Mauricio Macri came into office as the country's new president (2015–2019). The new administration brought a series of significant changes within the state, among them the reduction and dismanteling of certain areas and their technical teams.

are responsible for their forest resources,[12] but also, and more importantly, because in order to achieve legitimacy, the PICs had to be approved by provincial state agencies. In practice, this meant that PICs were presented and approved by the local Forest Administrations and other provincial state agencies, selected according to the criteria and demands imposed by each provincial government. In the words of a project Coordinator at the provincial level: "Each provincial government had to decide which specific governmental agencies were to oversee the plans. However, what was not up for debate was that in each province [Chaco, Salta and Santiago del Estero] [the plans] necessarily had to pass through the provincial government." Once the intervening provincial agencies were identified, formulators, technicians and national and provincial officials defined the necessary procedures and mechanisms, as well as the documentation to be presented for the approval of the PICs that went through three stages, in the following order: technical, legal, and environmental approval. The fact that the project worked with communities with precarious land tenure generated excessively bureaucratic processes to certify the right to use resources and land possession rights.

The experience and accomplishments in the tasks developed by the provincial agencies were central aspects of the project. When the Forest Law eventually adopted the PICs as forest plans, the provinces were prepared to carry them forward, based on this previous experience and accomplishments. Members of the executing team at the national level mentioned several of these aspects during an interview in 2019:

> We approached the topic acknowledging the province's status upon its forests. And acknowledging our role, [but] not as an executing unit; often, projects with external functions are like state parasites (…) they might be able to execute with the best of results, but this hardly ever translates into strengthening state capacities. And so, from the very beginning this project was conceived the other way around. (…) According to the national government, the project's main goal was the strengthening of the law. And to our team it was strengthening land tenure. Strengthening land tenure through the Forest Law. This was a key feature of our technical implementation.

These actors see the process of strengthening land tenure as a set of actions aimed at achieving provincial state recognition of indigenous and peasant's rights to use forest resources and possess land. Strengthening of tenure was not seen as a path to access a lands real estate market nor understood as a way of accessing to land titles. Instead, it was envisioned as a set of procedures, practices, and routines that, mobilized by certain provincial agencies, could enable the acknowledgment and legitimization of indigenous and peasant's rights. Therefore, national-level government officials and technicians deemed it crucial to involve provincial agencies in the project's validation. According to their understanding, the practices and mechanisms generated by the project would pave the way for the development of management plans devoted specifically to these sectors. Once the project had

---

[12]In a federal government system, the provincial level of government carries out most of the actions with a potential impact on forests and people who live in forest areas.

concluded, these plans could be easily continued through the approval of future annual operating plans. But also, because involving the provincial agencies was perceived as a way of engaging them and demanding that they participate in the policy of recognition, distribution, and access to land and resources.

3 State Processes of Demarcation and Certification of Indigenous and Peasant Lands

For the group of project formulators and technicians at the national level, assuming the strengthening of land tenure as a central goal of forest policy entailed negotiating and reaching agreements with the provincial governments, who established their own criteria and made adjustments. That is to say, the provincial governments and agencies continued to shape the project, laying the foundations, and defining the constraints regarding the type of tenure required to qualify as a recipient. As expressed by a national-level technician, some officials and technicians of the provincial agencies envisioned and understood "the implications of how you accept a project" and from then on began to impose constraints through which to regulate its scope.[13] Hence, the project's "scope" and "restrictions" had to do with the intention and outreach capability that each province wanted to give it according to political decisions. In Salta, the local government defined that the project would only target "communities" (indigenous and *criollo* families) that had perfect titles, i.e., titled lands.[14] As recalled by a member of the technical development team, "The provincial government told us:—'Don't bring me plans where you don't have titles.'" This shows the importance of plans as a means of legitimizing land possession and natural resource use (See Project Appraisal Document, p. 29). In contrast, in the province of Chaco, the government agreed to develop the project in "communities" located on public lands. On its part, the government of Santiago del Estero agreed to work with the figure of "holders with an owner's interest." The "will of the province" of Santiago del Estero, and of the Forest Administration in particular, favored that the indigenous and peasant communities that claim territories and have precarious tenures in departments with a great deal of territorial conflict could access to the project, in spite of all this. The fact that in Santiago del Estero, inhabitants recognize themselves as holders with an owner's interest, while in Chaco, they are holders of public land who demand land from the province,

---

[13]It should be noted that the provinces differ from one another in terms of the situation of indigenous and peasant communities. These differences have to do with legal recognition, ethnogenesis processes, forms of land tenure and resource administration, implementation of land survey policies, etc. The provinces have been responding differently to land regularization processes.

[14]Many indigenous communities in Salta inhabit private lands that are currently in use by the owner. Based on this situation, the government of Salta decided not to attest the peaceful occupancy, an eligibility criterion imposed by the World Bank to be a beneficiary of the project.

reveals the different relationships that indigenous people and peasants establish with each provincial state. In the latter case, they recognize the state as the owner of the land, whereas in the former case, possessors recognize themselves as landowners and seek through different means that the state gives proof of this situation. The fact that the province of Chaco has a land-granting policy, while Santiago del Estero and Salta do not, also explains the different positions of provincial governments. In the specific case of Santiago del Estero, this province is characterized by an agrarian structure with a significant presence of peasants, by the persistence of agricultural operations without defined boundaries, by the presence of landholders with an owner's interest but without property titles and unclear situations regarding ownership titles (Jara and Paz 2013). A large majority of the peasants' production units have not been registered.

The adjustments at a provincial level were carried out on the basis of a framework of guidelines and definitions pre-established as from the project's development, which defines a range of criteria to be fulfilled by the "communities" aspiring to become "recipients" of the forestry policy. Some of these requirements were imposed by the World Bank and the project refers to them as "eligibility criteria" and classifies them into two types: biophysical and social.[15] Two of these criteria are of particular interest in light of this paper's objectives. Among the biophysical criteria, one of them establishes that the community should have a minimum area (2000 ha) and a maximum area (100,000 ha) under community planning. Among the social criteria, one of them establishes as a requisite the existence of a situation of "peaceful and uninterrupted occupancy of the territory" by the community for more than 10 years (Integrated Plans Manual, p. 16).

> Among the eligibility criteria... there are a minimum and maximum expected number of families. Eventually, the maximum number exploded and there are many more. There is a minimum of territory that we also estimated based on Chaco's agrarian structure, of 20 families with at least 200 hectares. And we also set territory eligibility criteria. The [World] Bank demanded a criterion of peaceful and uninterrupted occupancy for more than 10 years. So, we had to find provincial instruments that could certify peaceful occupancy, especially taking into account occupancy that is not acknowledged by a property title and that could be contested in some way by a provincial State agency, when you bring the project to their offices (Interview with a member of the team of technicians and formulators at a national level, 2019).

---

[15]The biophysical criteria establish that (1) communities must be located in certain departments: San Martín and Rivadavia (Salta), Alberdi, Pellegrini and Copo (Santiago del Estero) and General Güemes (Chaco); (2) there must be a "customary use," by the members of a community, of an area of native forest belonging to any of the three categories of the Territorial Planning of Native Forests (Law 26,331); (3) that the minimum percentage of forest cover must be 60%. The project's social eligibility criteria establish that (1) communities must show their "interest in community and forest management projects" and (2) that their collective interest in participating must be "adequately documented" in a "letter of intent" (PIC Manual), (16–17); (3) that only those belonging to one of the following groups will be eligible: "indigenous communities," "peasant communities," and/or "small farmer and family farming groups"; (4) that the "community or governance unit" must be composed of a minimum of 10 and a maximum of 60 families (Integrated Plans Manual, p. 16); (5) that the application for investments must be of a community type.

Since many communities have neither demarcated nor certified their lands-territories, nor do they entirely know the area they occupy, provincial state agencies carried out land survey and delimitation work. Some provinces, such as Santiago del Estero, created the Registry of Applicants for the Regularization of Land Tenure (better known as the Registry of Holders) in 2006 to this end. Although the Registry has undertaken a series of tasks associated with land regularization, registering 1000 peasant families in an area of approximately 150,000 ha located in different departments of the province (De Dios 2012), the lack of funding of the Registry in recent years, among other factors, hindered progress with the regularization plan.

As for the criterion of peaceful occupancy, this mainly meant demonstrating the absence of conflict in the delimited territorial area. Among the ways of verifying this, field visits were made to collect first-hand information through surveys and to issue reports on the peaceful occupancy or absence of conflict. These visits were used to determine whether or not legal action had been taken on the delineated areas.

However, beyond the presentation of these criteria as objective and quantifiable data, the truth is that in both cases the concerned communities, along with the technicians of the rural development agencies that advised them, have manipulated the actual data in various ways and have developed different strategies in order to become recipients. For example, in certain cases when the community did not reach the minimum area (2000 ha) under planning required by the project, they decided to join another community and present themselves as a single recipient community. In some cases, communities whose lands were in conflict with third parties came forward, but under the strategy of leaving those conflictive areas aside when submitting the Integrated Community Plans. "If there was legal action in an area," a technical expert from Santiago del Estero told me, "it would be dismissed."

The provincial agencies involved in the project validated processes by issuing reports, administrative records, resolutions, and notes, surveying and drawing up territorial boundaries. This is especially important given that the project is being developed in many lands that had not been previously surveyed, delimited, or recognized by provincial and/or national government agencies. Land demarcation is essential to avoid the overlapping of PIC-certified management units.[16] The PIC is both the means by which communities establish their territorial boundaries and the process by which state agencies acknowledge the territories used by these communities. The acknowledgment of use is not a synonym of a legal recognition of

---

[16]The project uses the PIC as a tool to collect proposals from communities on activities associated with the forest: protection, restoration, management, and use of forest resources. As the PIC operates and makes investments in the forest, it was conceptualized as a management plan. Its development should include the development of a Sustainable Management Plan (SMP) for the Forest, through which the execution of works and activities can be justified (...) and the negative impacts of other productive activities (...) on the forest reduced" (Integrated Community Plans Manual, p. 19). In each provincial jurisdiction, the SMP must be approved by the local enforcement authority of National Law 26,331.

ownership. Rather, it is only a precedent that might be useful in the future to substantiate such ownership by the communities. Part of this territorial demarcation that is included in the PIC involves surveying and clearly determining the occupied territory, and establishing consensus and guidelines for the use of community space: of its native forests, pastures, shrubs, trees, wells, watercourses, wild fauna, firewood, community infrastructure, etc. One of the intentions of this territorial delimitation is to produce information that allows avoiding possible conflicts with neighboring communities due to potential overlapping, especially for communities with no defined boundaries.

State recognition of territories involves implementing various procedures and practices, such as registering different types of holders, certifying the proposed territorial units, and drawing up of maps, reports, and administrative records. To describe and assess the implications of demarcation and State recognition, a nationwide technician compared and differentiated the project with a State policy for the mapping of indigenous lands, the Territorial Survey of Indigenous Community Program (Re.Te.Ci for its Spanish acronym):

> This is not a survey like Re.Te.Ci where people did not know the extent of the survey and did not have a copy of the map approved by anyone. Through this project, communities obtain an administrative record that certifies their right to the use of resources and their right of possession over the land (2019).

The surveys, delimitations and certifications of administrative and land possession acts carried out within the framework of the PICs are some of the actions that were implemented as part of the territorial planning that came hand in hand with the strengthening of land tenure. In Santiago del Estero two agencies at the provincial level are working in coordination and articulation with the Forest Administration and the Environment Administration: the Emergency Committee and the Registry of Holders, both of which report to the Administration of Inter-Institutional Relations. The Committee was created in 2007 to address land conflicts and the Registry of Holders was created to regularize land tenure of peasant families. The Registry of Holders "was conceived as part of a policy aimed at compensating inequalities, so that holders would achieve better conditions or greater possibilities of access to justice" (De Dios 2012: 122). This agency conducts a series of tasks, such as field visits for the registration of peasant families[17] as holders with an owner's interest, and provides economic support through subsidies to access legal representation for the twenty-year statute of limitations on land ownership.[18] It also technically assists in the territorial delimitation and the drafting

---

[17]Most of the province's holders are located on private lands and not on public lands (De Dios 2012).

[18]In Argentine Chaco many peasant families carry out production activities on their fields recognizing themselves as sole owners of these lands. They are holders with the intention of an owner. This possession generates rights and, as per the Civil Code, if it continues for twenty uninterrupted years, possession can give way to ownership with the certifying title. Thus, the possessor earns the right to ownership and after twenty years can become the owner of the land he inhabits.

of territorial survey plans[19] containing information on the territorial area to be expropriated and handed over (De Dios 2012). Its actions and, fundamentally, the documentation issued by the agency are seen as proof of the presence and occupancy of the settler in the event of a lawsuit or an eviction order by a third party. The following testimony illustrates of the value of certain actions and procedures that produce and delimit territorialities:

> In Santiago we can reach a certain limit where the province by means of these instruments and agencies certifies the management plan and people receive a map, an administrative act. Then, if a grabber comes in tomorrow and wants to buy this land, people have a way of proving that they are former inhabitants. So, that is why we talk about strengthening tenure through the law's management plans. Because the law is not designed for that which we are taking advantage of (Interview with a nationwide developer and official, 2019).

In turn, the Emergency Committee's main task is to intervene in conflict situations, where "alleged or real title holders drive bulldozers over the peasants' possessions (…); or when they intend to extract forest resources without their consent; or when the local police fail to hear holders' complaints, (…)" (De Dios 2012: 123). The Committee responds to this type of situations, informing and advising on the possibilities of legal defense. It verifies and certifies by issuing reports on forms of peaceful occupancy or conflict in a given territorial unit. As one interviewee put it, "The Committee rules on the site delimited by the PIC. It issues an opinion describing whether there is conflict. They tell you if it's all right there, if they're traditional occupiers. Then, later on, the province approves the plan." As it is clear from the testimony, the tasks carried out by these agencies are supplementary, because while the Emergency Committee deals with situations of confrontation and dispute and issues a ruling on the existence or not of conflicts in a certain place, the Registry of Holders carries out the registration of the peasant families and technically aids in the execution of a territorial survey for the subsequent drawing up of a map.[20] Their actions are highly valued as steps toward a "cadastral legal system." As one formulator and technician at the national level stated during an interview in 2019:

> In Santiago del Estero the provincial government provides support to the development of the projects by making available its instruments, such as the Emergency Committee and the Registry of Holders. Why? Because the truth is that the project is also contributing to somehow generate a cadastral legal system of a territory that the local government did not know how to approach. We asked [the provincial government to] define the territory, to tell us which forests they were going to manage... So, this community agrees with this other

---

[19]Once prepared, land survey plans must be approved by the provincial Cadaster Administration.

[20]As part of the institutional architecture of the project, a provincial Advisory Council was set up that grouped together agencies from different areas—lands, forests, environment, indigenous and peasant affairs—and also summoned local peasant and indigenous organizations. The role played by the advisory council is central, since it has the capacity to design a strategy for access to the territories and to carry out the first tentative evaluation on the feasibility of the projects presented. The Council is a provincial work table where the Forest Project discusses with the authorities of the multiple areas in order to settle and guide on different aspects of the life of the project.

community on the boundaries and then the Emergency Committee certifies this and the territory is gradually organized a bit, a territory that in the past was mostly just undefined patches. The provincial government supports what we call the strengthening of land tenure, especially in situations where the territories of communities had not been surveyed by [State] agencies.

Although the executing unit [at the national level] and the provincial State agencies were aware of the existence of communities classified as "peaceful" versus others classified as "conflictive," they made the decision to not "avoid territories with conflicts." Therefore, the dissemination of the project reached all communities. However, in the case of the communities with conflicts, they assessed the community's situation, when the conflict had begun, whether some kind of resolution had been reached and whether the whole community territory, or only a part of it, was affected by the conflict. In some cases, when there were conflicts within the community territories, it was decided to "disengage those areas" and work only with areas that were free of conflict.

Between the provincial executing unit and the Emergency Committee, they designed a strategy to gather information that would allow them to assess the degree of conflict.[21] They arranged for training on land, territoriality, and rights, with the participation of community members and technical staff from the project's rural development agencies. During the training delivered by the Committee's technical specialists, the exchanges were geared toward extracting data on the land situation. The members of the local executing unit positively evaluated the data collection methodology and the role played by the Committee, since they enabled a rapid evaluation process on the territorial situation and allowed access to the report in a short period of time.

During the process, both the Committee and the Registry of Holders gradually incorporated bureaucratic and administrative practices and routines that facilitated the coordination of procedures and guarantees that are part of the life and operation of the project.

4   Final Considerations

By analyzing different stages and aspects of the implementation of a forestry policy and focusing on bureaucratic practices and State procedures, I highlighted the ways in which the State and its agents organize, classify, demarcate and produce new territories and territorialities. I therefore showed how the State, through the Forest project, deploys means of regulation and intervention that reinforce its authority and legitimacy over natural resources, territories and the people who inhabit them. The

---

[21]This did not mean that they accepted the criterion of conflict as something given, but rather the opposite. As one of my interviewees stated, demonstrating or not demonstrating conflict is an ambiguous criterion, since areas or communities without conflict could cease to remain as such at some point during the implementation of the project.

analysis of documentation is key to understanding how the State and its agents speak to us; documents are fundamental devices for the legitimization of their own authority and superiority. Through its practices and procedures, the project has managed to consolidate and strengthen the power of the State in its functions of surveying, organizing, demarcating, validating and recognizing certain forms of land occupation and tenure by the indigenous and peasant populations. In this sense, its project to strengthen and regularize land tenure can be seen as an extension of the dominion over the territories and their inhabitants that the Nation-State aims at. The analysis unveils how the project contributes to certain fundamental State objectives, such as the planning and demarcation of land and territory. I also pointed out how the funding agency imposes its conditions on the State, establishing "eligibility criteria" that restrict the scope of what can be surveyed, demarcated and organized. The study shows how funding agencies influence and determine the ways in which states categorize, classify, and rank the subjects and lands eligible for this public policy. Hence, the study consists not only of a critical presentation of the project's scope, ideals and values, but also, and fundamentally, of an analysis of its concrete application.

Rather than thinking about a disjointed State and actions, the study offers empirical evidence about the way in which the Native Forest project imposes a set of bureaucratic procedures and practices that are instituted and naturalized as the only possible method to organize and acknowledge the lands-territories of peasants and indigenous people. The study analyzes the different procedures and practices that the State uses to legitimize its actions and dominion, and the capacity of its agents to present them as natural, ensuring that they are not questioned or resisted by local actors. The State procedures deployed during the forestry policy are part of what Ferguson and Gupta (2002) call "framing" to refer to how the State makes use of them to impose a certain order that comes to be taken as natural. The above analysis shows how behind the rhetoric of strengthening land tenure there are assumptions and ways of organizing and recognizing lands and territories that match state-authorized and validated forms.

References

Aguiar S et al (2018) ¿Cuál es la situación de la Ley de Bosques en la Región Chaqueña a diez años de su sanción? Revisar su pasado para discutir su futuro. Ecología Austral 28:400–417

Archivo General de la Nación (AGN) (2017) Informe de Auditoría

Cabrol D, Cáceres DM (2017) Las disputas por los bienes comunes y su impacto en la apropiación de servicios ecosistémicos: La Ley de Protección de Bosques Nativos, en la provincia de Córdoba Argentina. Ecol Austral 27(1):134–145

Camba Sanz G, Aguiar S, Vallejos M, Paruelo M (2018) Assessing the effectiviness of a land zoning policy in the Dry Chaco. The case of Santiago del Estero, Argentina. Land Use Policy 70:313–321

Cotula L (2012) The international political economy of the global land rush: a critical appraisal of trends, scale, geography and drivers. J Peasant Stud 39(3–4):649–680

Das V, Poole D (2008) El Estado y sus márgenes. Etnografías comparadas. Cuadernos De
    Antropología Social 27:19–52
De Dios R (2012) Ordenamiento territorial e inclusión social en Santiago del Estero. Revista
    Realidad Económica 268:115–127
Di Paola ME, Ramírez S (2014) Fortalecimiento de la tenencia de la tierra. Proyecto sobre Bosques
    Nativos y Comunidad (P132846) FAO-BM. Buenos Aires
Ferguson J, Gupta A (2002) Spatializing states: towards an ethnography of neoliberal
    governmentality. Am Ethnol 24(29):981–1002
Grau HR, Gasparri NI, Aide TM (2005) Agriculture expansion and deforestation in seasonally dry
    forests of north-west Argentina. Environ Conserv 32(2):140–148
Hansen MC, Potapov PV, Moore R, Hancher M, Turubanova S, Tyukavina A, Thau D (2013)
    High resolution global maps of 21st-century forest cover change. Science 342(6160):850–853
Informe de estado de Implementación 2010–2015, Ministerio de Ambiente y Desarrollo
    Sustentable, 2015
Informe Productivo Provincial, Chaco (2019) Subsecretaría de Programación Microeconómica,
    Secretaría de Política Económica, Ministerio de Hacienda de la Nación, Presidencia de la
    Nación
Informe Productivo Provincial, Santiago del Estero (2019) Subsecretaría de Programación
    Microeconómica, Secretaría de Política Económica, Ministerio de Hacienda de la Nación,
    Presidencia de la Nación
Informes Productivos Provinciales Salta (2017) Año 2, 12. Secretaría de Política Económica,
    Subsecretaria de Programación Microeconómica, Dirección Nacional de Planificación Sectorial
    y Dirección Nacional de Planificación Regional. Ministerio de Hacienda, Presidencia de la
    Nación
Jara C, Paz R (2013) Ordenar el territorio para detener el acaparamiento mundial de tierras.
    Conflictividad de la estructura agraria de Santiago del Estero y el papel del Estado. Proyección
    XV:171–195
Lefebvre H (1974) La producción del espacio social. Revista De Sociología 3:219–229
Nolte C, le Polain de Waroux J, Munger T, Reis N, Lambin EF (2017) Conditions influencing the
    adoption of effective anti-deforestation politics in South America's commodity frontiers.
    Global Environ Change 43:1–14
Manual Planes Integrales Comunitarios. Proyecto "Bosques Nativos y Comunidad" (BIRF
    8493-AR/PNUD ARG 15/004), Ministerio de Ambiente y Desarrollo Sustentable, presidencia
    de la Nación
Marco de Gestión Ambiental. Proyecto "Bosques Nativos y Comunidad". Secretaría de Ambiente
    y Desarrollo Sustentable (2015)
Marco Integral Comunitario. Proyecto "Bosques Nativos y Comunidad. Secretaría de Ambiente y
    Desarrollo Sustentable-Jefatura de Gabinete de Ministros"
Marcus G (1995) Ethnography in/of the world system: the emergence of multi-sited ethnography.
    Annu Rev Anthropol 24
Observatorio de Políticas Públicas (2012) Comunidades Aborígenes y Ordenamiento Territorial.
    CAT. OPP/CAG/2012–02. Cuerpo de Administradores Gubernamentales, Buenos Aires
Stoler AL (2010) Archivos coloniales y el arte de gobernar. Revista Colombiana De Antropología
    2(46):465–496
Torella S, Piquer-Rodríguez M, Levers C, Ginzbrg R, Gavier-Pizarro G, Kuemmerle T (2018)
    Multiscale spatial planning to maintain forest connectivity in the Argentine Chaco in face of
    deforestation. Ecol Soc 23(4):37
The World Bank. International Bank for Reconstruction and Development. Project Appraisal
    Document (2015, March 6)
The World Bank. Restructuring Paper on a Proposed Project Restructuring of Forests and
    Community (P132846) (2019)

Tsing AL (2005) Friction. An ethnography of global connection. Princeton University Press, Princeton

Venencia CD, Correa JJ, Buliubasich C, Seghezzo L (2012) Conflictos de tenencia de la tierra y la sustentabilidad del uso del territorio del Chaco salteño. Avances En Energías Renovables y Medio Ambiente 16(1):105–112

Annexes

Sebastián Valverde

Annex I. Notes on the Academic Report on the Impact of COVID-19 on Indigenous Peoples

Abstract The following text is a summary of the most important passages—and adapted for the publication of this book—of the *Observatorio Universitario de Buenos Aires (OUBA) Nº13* report titled "*Quinientos años no es nada*", published on July 10, 2020, dedicated to the *Informe ampliado: efectos socioeconómicos y culturales de la pandemia COVID-19 y del aislamiento social, preventivo y obligatorio en los Pueblos Indígenas del país -Segunda etapa- Junio 2020.* Abeledo et al. (2020). *Informe ampliado: efectos socioeconómicos y culturales de la pandemia COVID-19 y del aislamiento social, preventivo y obligatorio en los Pueblos Indígenas en Argentina-Segunda etapa, junio 2020* (http://antropologia.institutos. filo.uba.ar/sites/antropologia.institutos.filo.uba.ar/files/info_covid_2daEtapa.pdf).

As it is pointed out on the Web site, the OUBA's objective is to "(…) make visible, from an academic point of view, the nodal spaces of the politics of the city of Buenos Aires where the State has retreated from its designated role or has reoriented itself towards other kinds of actions, generating situations that intensify inequality in the access to fundamental rights" http://general.filo.uba.ar/ observatorio-universitario-de-buenos-aires-ouba.

S. Valverde
Consejo Nacional de Investigaciones Científicas y Técnicas (CONICET), Universidad Nacional de Luján, Luján, Argentina

Consejo Nacional de Investigaciones Científicas y Técnicas (CONICET), Facultad de Filosofía y Letras, Universidad de Buenos Aires, Buenos Aires, Argentina

C. M. Minaverry and S. Valverde (eds.), *Ecosystem and Cultural Services*, The Latin American Studies Book Series, https://doi.org/10.1007/978-3-030-78378-5

101

Keywords COVID-19 • Indigenous communities • Social aspects

Introduction

As it is already known, the current pandemic COVID-19 is having an profound impact onvarious spheres of society's life, provoking a deep transformation of all daily social relations. The ASPO, the acronym for the obligatory and preventive social distancing,[1] enacted by the Argentinian National Government as from March 20th of 2020, constituted a measure to avoid the spread of the COVID-19 virus, avoiding a much higher number of infections and deaths. The immediate consequence has been the deceleration of employment and an abrupt income retraction of large sections of the country and, of course, also of members of indigenous peoples —largely informally employed—having a radical influence on their community's economy.

Nevertheless, and above all in this chapter, we aim to make visible the socioeconomic inequalities, stigmatization and, in some occasions, the criminalization associated with their sociocultural condition of indigenous people, that have been exacerbated in these last months with specific instances of institutional violence. Also, we will mention the national government's reparation policies,[2] the already existing[3] and the community support networks in the middle of this crisis. Without these social bonds and the state policies, the social consequences would be devastating.

The Impact of the Pandemia in the Economic Life of Indigenous Peoples

It is important to highlight that the current context implies a sharp disparity, between those who count on a steady income by means of a salary, and those who do not, as is the case of the majority of indigenous women, who perceive an informal income. In relation to this, Sebastián Valverde[4] pointed out:

---

[1] In Spanish: "Aislamiento Social, Preventivo y Obligatorio".
[2] By this we mean for example the IFE or the Tarjeta Alimentar. These were policies implemented by the national government when the pandemic began in March 2020. The Tareta Alimentar was implemented during the first days of the presidency of Alberto Fernández (who took office in December 2019 and will remain in office until 2023).
[3] We are considering for example retirement pensions and other kinds of pensions, such as AUH.
[4] Sebastián Valverde is Professor of Economic Anthropology, who has been working on this topic specifically with indigenous peoples.

"If according to official figures the labor informality for [Argentine] society as a whole is at 44%,[5] among members of indigenous peoples it can get up to 70%, 80% and 90% or practically the entirety of the workers belonging to an indigenous people in certain contexts in some regions or areas in particular"[6] (Observatorio Universitario de Buenos Aires 2020: 2).

The overwhelming loss of income that has resulted from the halt or drastic decrease of a significant part of activities has affected severely the entirety of the indigenous peoples. Valverde stressed on this respect:

"In this sense, it is important to point out that members of these groups hold low-skilled jobs in rural settings or, in the cities, wage labor that is severely affected in the current context: women as domestic staff and men as construction workers"[7] (Observatorio Universitario de Buenos Aires 2020: 2).

The resources arranged by the national government were key, at least as a mitigating measure, by granting income, as that derived from the "Alimentar" programme, the AUH,[8] receiving social benefits, food cards and the recent IFE,[9] among other contingency plans, of wide coverage for millions of Argentines. Alejandro Balazote[10] indicated about this:

"It would be desirable that the new government would begin to rethink –maybe 'inaugurate'- new approaches in the relationship that the Nation-State has historically had with the indigenous peoples. This implies discussing and taking a clear stance with regards to a historiography that has systematically denied the presence, and more so the genocide of indigenous peoples in our country"[11] (Observatorio Universitario de Buenos Aires 2020: 2).

Balazote (Observatorio Universitario de Buenos Aires 2020) concluded stressing that a public agenda has to be designed that entails a historical reparation, and that it certainly has concrete consequences in relation to access to land and natural resources.

---

[5]Based on data from Bertranou and Casanova, 2014, derived from the *Censo Nacional de Población y Vivienda de 2010*, see Informe Abeledo et al. (2020).

[6]Original in Spanish: "Si en el conjunto de la sociedad la informalidad es del 44% (de acuerdo a datos oficiales), en los integrantes de los pueblos indígenas puede alcanzar en algunas regiones o zonas en particular el 70%, 80% y 90% o prácticamente la totalidad de los trabajadores pertenecientes a los pueblos originarios en determinados contextos".

[7]Original in Spanish: "En este sentido, cabe destacar la inserción de los integrantes de estos grupos en tareas de menor calificación en los ámbitos rurales o bien, en las ciudades, como asalariados en empleos que se ven severamente afectados en el actual contexto: en el servicio doméstico las mujeres y en la construcción, los hombres".

[8]In Spanish, this acronym stands for Universal Allowance per Child.

[9]In Spanish, this acronym stands for Emergency Family Income.

[10]Doctor in Anthropology, Postgraduate Secretary and full Professor of the Rural Anthropology seminar of the Facultad de Filosofía y Letras (UBA).

[11]Original in Spanish: "Sería deseable que el nuevo gobierno comience a repensar –quizás 'inaugurar'—nuevos formatos en el vínculo que históricamente ha tenido el Estado-Nación con los pueblos originarios. Esto implica discutir y tomar una posición clara frente a una historiografía que sistemáticamente ha negado la presencia misma, y desde ya el genocidio de los pueblos indígenas de nuestro país".

The Presentation of an Academic Report About the Situation of Indigenous People During the COVID-19 Scene

On July 10, 2020, a hundred members of different sectors of academia of Argentina publicly presented two different reports: 1. the main achievement of the "Efectos Socioeconómicos y culturales de la pandemia" and "Aislamiento Social Preventivo y Obligatorio (ASPO) en pueblos indígenas".[12] Specialists from thirty research groups belonging to twelve different universities and/or research units of CONICET worked together, who despite the limitations, consequence of the ASPO, were able to survey comprehensively close to 80% of the indigenous peoples.

The compiled information covered various indigenous peoples in Argentina: toba-qom, mbya moqoit, guaraní, avá guaraní, kolla, diaguita, diaguita-calchaquí, wichí, huarpe, quechua, aymara, nivaclé, tonokote, omaguaca, tastil, günün a küna, comechingón, comechingón-camiare, ocloya, iogy, chané, tapiete, sanavirón, ranquel, wehnayek, atacama, lule, quilmes, pehuenche, mapuche, mapuche-pehuenche, tehuelche, mapuche-tehuelche, selk'nam, haush and selk'nam-haush. Geographically, the project covered all the regions of Argentina: Área Metropolitana de Buenos Aires, Pampeana-Centro, Noroeste, Noreste, Cuyo and Patagonia (Figs. 1 and 2).

Indigenous Communities in Largest Urban Agglomerations

Currently, most of the members of indigenous peoples in Argentina live in cities. That is why one of the main issues since March 2020 has been that of members of indigenous peoples in urban settings in conditions of poverty and with varying degrees of infrastructure deficiency. Juan Engelman[13]explained (Fig. 3):

> "As from the constitutional amendment of 1994 the legal rights of indigenous peoples, as legal subjects, have been broadened and they were able to request legal representation, in administrative terms, as indigenous communities. Safety and support nets diversified. Especially in urban and peri-urban areas, community leaders, many times integrated to neighborhood political parties, have acquired experience before state bureaucratic bodies

---

[12]The project has been coordinated by Sebastian Valverde. See: http://antropologia.institutos.filo. uba.ar/sites/antropologia.institutos.filo.uba.ar/files/info_covid_2daEtapa.pdf.

[13]Researcher at the CONICET and at the Instituto de Ciencias Antropológicas (ICA) de la FFyL de la UBA and lecturer at the UBA and at the UNLu.

Fig. 1 Pachamama ceremony in a day of territorial claims in El Barrio de Kanmar, town of Glew, Almirante Brown County, Buenos Aires province, January 2017. *Source* Sebastián Valverde

and other types to take identity, class and land claims forward. At present they are not passive subjects, but they are organized as a people"[14] (Observatorio Universitario de Buenos Aires 2020: 5–6).

Racism and Institutional Violence

The report warns about several incidents registered throughout the country. It emphasizes on the dynamic that has been occurring in the complex pandemic scenario, which is associated with the intensification and exacerbation of instances of racism, discrimination, verbal and physical violence towards members of

---

[14]Original in Spanish: "A partir de la reforma constitucional de 1994 se ampliaron los derechos legales de los pueblos originarios como sujetos de derecho y pudieron tramitar personerías jurídicas como comunidades indígenas en términos administrativos. Se diversificaron redes de contención y asistencia comunitaria. En especial en zonas urbanas y periurbanas las y los referentes, muchas veces integrados en partidos políticos barriales, adquirieron creciente experiencia ante instancias burocráticas estatales y de todo tipo para llevar adelante por igual reclamos de identidad, de clase y territoriales. En la actualidad no son sujetos pasivos sino que se organizan como pueblo".

Fig. 2 Leader Nala Clara Romero of the Lma Iacia Qom community, San Pedro, Province of Buenos Aires. *Source* Sebastián Valverde

Fig. 3 Members of the "Consejo Indígena de Almirante Brown" and of the "Coordinación de Pueblos Originarios de Almirante Brown" gathered because of a territorial claim with government officials of the Almirante Brown municipality, the Consejo de Participación Indígena of the province of Buenos Aires and the Instituto Nacional de Asuntos Indígenas. June 2018. *Source* Sebastián Valverde

indigenous peoples, through arbitrary actions, and/or serious abuses perpetrated by officials of different public bodies, healthcare institutions and/or law enforcement bodies, adopting, in some cases, deeply traumatic and conflictive characteristics.[15] On this respect, Ana Carolina Hecht[16] stated:

> "Racism and discrimination towards indigenous communities have come back in the province of Chaco due to a disease that has spread a lot and arrived to the area thanks to wealthy travelers of the upper classes that came back from Europe. This seems to have been forgotten immediately with the accusations of 'infected indians' and the closing down with roadblocks of the accesses to the stigmatized qom communities by other local inhabitants"[17] (Observatorio Universitario de Buenos Aires 2020: 7).

Growing Problems in the Healthcare Sector

The pandemic also visibilized profound historical inequality in the access to basic rights, especially regarding health care. Dr. Tujuayliya Gea Zamora, physician of Wichi origin, explained in the recent presentation "Racism and the Pandemic",[18] that:

> "It is necessary to highlight that this epidemiological profile is derived from decades of dispossession and forced land evictions, and a profound historical inequality in the access to basic rights as formal employment, water and education"[19] (Observatorio Universitario de Buenos Aires 2020: 10).

As Dr. Gloria Mancinelli[20] stated, the communities themselves have come out to report not only the lack of consultation but also the necessity of participation, of working together with the communities in setting the scope of the measures to be considered in the local contexts. On this respect, she also added that:

---

[15]Original in Spanish: "Una dinámica que se viene dando en este complejo escenario de pandemia, se asocia con la profundización y exacerbación de situaciones de racismo, discriminación, violencia verbal y física hacia los integrantes de los pueblos originarios, a través de acciones arbitrarias, y/o graves abusos por parte de funcionarios de diversos organismos públicos, instituciones sanitarias y/o fuerzas de seguridad, asumiendo en algunos casos características sumamente conflictivas y traumáticas".

[16]Doctor in Anthropology of the FFyL of the UBA and a specialist in indigenous communities of Chaco.

[17]Original in Spanish: "El racismo y la discriminación hacia las comunidades originarias volvieron a decir presente en la provincia de Chaco debido a una enfermedad que se propagó mucho y llegó por viajeras de clase acomodada que volvieron de Europa. Esto parece haberse olvidado de inmediato con las acusaciones de 'indios infectados' y el cierre con barricadas de los accesos a comunidades qom estigmatizadas por otros pobladores locales".

[18]Original in Spanish: "Racismo y Pandemia".

[19]Original in Spanish: "Es necesario remarcar que este perfil epidemiológico se deriva de décadas de despojos y desalojos territoriales y una profunda desigualdad histórica en el acceso a derechos básicos como el acceso al trabajo formal, el agua y a la educación".

[20]Professor of Anthropology at the UBA and member of the ICA.

"It is necessary to have in mind that the pressing conditions under which the different indigenous peoples find themselves and the lack of concrete action on behalf of the different levels of the State, forces them to take action [...] which come from different sectors, as biomedicine and local knowledge, from the own experience of the indigenous communities and organizations."[21] (Observatorio Universitario de Buenos Aires 2020: 11).

"500 Years to Subjugate Us, 500 Years to Liberate Us".[22] The History of Indigenous Peoples: A History of Resistance

"The report about the impact of Covid-19 on indigenous peoples entailed the presentation of an academic document published collectively, which diverges from the protocols that guide the flow of academic knowledge, and individualistic and competitive production generated by a certain coercion to publish[23]" (Observatorio Universitario de Buenos Aires 2020: 11). Liliana Tamagno[24] considered that this publication is vital for our discipline and hold that:

"The pandemic crisis crudely revealed the inequality in our society and the ways in which it affects indigenous peoples, making it clear that if this inequality is not reverted, the possibility to comply with rights, that are theirs by law, is completely blocked".[25]

Tamagno concluded that:

"It is important to make it clear that the reported situations should not contribute to us thinking of them [the indigenous peoples] only as victims. The history of resistance also denies the condition of vulnerability that is frequently attributed to them as an intrinsic

---

[21]Original in Spanish: "Las propias comunidades han salido a denunciar no solo la falta de consulta sino la necesidad de participación, de trabajar junto y con las comunidades en la definición de medidas a considerar en los contextos locales. Es necesario tener presente que las condiciones acuciantes en las que se encuentran los diversos pueblos indígenas y la falta de respuestas concretas por parte de los diferentes niveles del Estado, los lleva a tomar iniciativas propias (...) que provienen de diversos ámbitos, como la biomedicina y conocimientos locales, desde las experiencias propias de comunidades y organizaciones indígenas".

[22]"500 años para sojuzgarnos, 500 años para liberarnos [five hundred years to subdue us, five hundred years to liberate us]" is part of a historical slogan of indigenous peoples' movements of Latin America.

[23]The original quote was: "El informe sobre el impacto del COVID-19 en pueblos indígenas implicó algo vital para nuestra disciplina, una reacción en cascada con interpretaciones super-adoras de un culturalismo que no asume la necesidad de pensar la etnicidad en su articulación con la desigualdad propia de la sociedad de clase. Una forma de producción colectiva que se aleja de los protocolos que rigen la circulación del saber académico y de las producciones individualizantes y competitivas generadas por cierta coerción a publicar y que generaron lo que denominé "malestar en la etnografía"".

[24]Researcher at CONICET and director of the Laboratorio de Investigaciones en Antropología Social (LIAS) of Facultad de Ciencias Naturales y Museo UNLP.

[25]Original in Spanish: "La crisis de la pandemia reveló con crudeza la desigualdad de nuestra sociedad y los modos en que afecta a los pueblos indígenas, dejando en claro que si la desigualdad no se revierte, se ve totalmente obturada la posibilidad del cumplimiento de los derechos que, por ley, les asisten".

property, which implies another way of underestimation. The organizational forms, critical reflections, knowledge, dreams and utopias of indigenous peoples show a collective community behavior which has been reflected in many moments of this report and that, although subsumed in the logic of capital, opposes it; alternate ways of thinking and acting, representations and practices founded in the past, effective in the present and conditioning factors, of a future common to all"[26] (Observatorio Universitario de Buenos Aires 2020, p. 12).

Final Considerations

The idea is to move forward with a common agenda shared between the several public agencies, academics and indigenous organizations in the collective construction of knowledge and concrete public policy.

Ultimately and as the conclusion of the report points out: "The voice of the peoples and their ancestral knowledges play a fundamental role in these stages of history"[27] (Abeledo et al. 2020, p. 11).

The impact of this report in Argentina has been enormous and, and on August 14, 2020, it was presented to the highest national authorities responsible for the application of public policies on indigenous peoples. We hope that this report contributes to the undoing of centuries of racism and inequality, a situation that has regrettably worsened in the context of COVID-19.

References

Abeledo S, Acho E, Aljanati L, Aliata S, Aloi J et al (2020) Informe ampliado: efectos socioeconómicos y culturales de la pandemia COVID-19 y del aislamiento social, preventivo y obligatorio en los Pueblos Indígenas del país. Segunda etapa, June 2020. Last access: 9 Oct 2020. http://antropologia.institutos.filo.uba.ar/sites/antropologia.institutos.filo.uba.ar/files/info_covid_2daEtapa.pdf

Observatorio Universitario de Buenos Aires (2020) Informe OUBA N.º13. Quinientos años no es nada. Buenos Aires: Secretaría General, Facultad de Filosofía y letras, Universidad de Buenos Aires. Last access: 9 Oct http://novedades.filo.uba.ar/sites/novedades.filo.uba.ar/files/documentos/OUBA%20NRO.%2013%20DIFUSION%20FINAL.pdf

---

[26]Original in Spanish: "Es importante dejar sentado que las situaciones denunciadas no deben contribuir a que los pensemos sólo como víctimas. La historia de resistencias desmiente también la condición de vulnerabilidad que frecuentemente se les atribuye como propiedad intrínseca, lo que implica otro modo de subestimación. Las formas organizativas, las reflexiones críticas, los saberes, los sueños y las utopías de los pueblos indígenas muestran un comportamiento colectivo comunitario que se refleja en muchos momentos de este informe y que, aunque subsumido en la lógica del capital, se opone a él; formas alternas de pensar y actuar, representaciones y prácticas fundadas en el pasado, efectivas en el presente y condicionantes, de un futuro común a todos".

[27]Original in Spanish: "La voz de los pueblos y sus saberes ancestrales juegan un rol fundamental en estas instancias de la historia".

Annex II. Prejudice Towards Mapuche People: The Attribution of Foreignness as a Strategy for Stigmatization

Juan Carlos Radovich, Sasha Camila Cherñavsky, Nadia Molek, Rocío Miguez Palacio, Ailén Flores, Frank James Hopwood, and Sebastián Valverde

J. C. Radovich · S. C. Cherñavsky · N. Molek · R. M. Palacio
Facultad de Filosofía y Letras, Universidad de Buenos Aires, Buenos Aires, Argentina
e-mail: radovich@retina.ar

A. Flores
Social Sciences Department, Universidad Nacional de Luján, Argentina
e-mail: ailensflores@hotmail.com

F. J. Hopwood
University of Ghent, Ghent, Belgium

S. Valverde
Consejo Nacional de Investigaciones Científicas y Técnicas (CONICET), Universidad Nacional de Luján, Luján, Argentina

Consejo Nacional de Investigaciones Científicas y Técnicas (CONICET), Facultad de Filosofía y Letras, Universidad de Buenos Aires, Buenos Aires, Argentina

C. M. Minaverry and S. Valverde (eds.), *Ecosystem and Cultural Services*, The Latin
American Studies Book Series, https://doi.org/10.1007/978-3-030-78378-5

Introduction[28]

Historically, certain social groups have invented and deployed a prejudice towards the Mapuche[29] people, which states that the Mapuche constitute a supposed "Chilean invader" that have invaded the "real Argentinian indigenous people", meaning by this the Tehuelche ("Aónikenk") people and others (e.g., the "Poyas"). This extremely fallacious explanation has already been largely refuted by empirical data gathered by disciplines such as history, anthropology and archeology. Nevertheless, the "Chilean invader" label, which is being replicated over and over again by certain social discourses and mass media, in education as well as in political spheres, is still working as a highly functional and consolidated statement that legitimizes particular interests. In fact, this prejudice has consolidated and strengthened the discrimination, racism and criminalization towards the Mapuche people in the Argentinian society (Radovich et al. 2014; Balazote et al. 2014).

The prejudice emerged in the late nineteenth century, in the context of the construction of the Argentinean Nation-State. The Argentine statesmen evaluated the need of consolidating the border with Chile, and consequently, the border conflicts between Argentina and Chile were accentuated. On the other hand, the need of "homogenizing" and "Argentinizing" the Norpatagonic region population took the State to advance over the territories occupied by the Mapuche people and to shape intentionally the image of the original Mapuche people as an alleged foreign "invader", who supposedly would have eliminated the "true" Argentine indigenous people: the Tehuelche.

The establishment of the political border between Argentina and Chile must be related to the construction of a sociocultural barrier, a border selfing and othering

---

[28]This paper is an updated and adapted version by Eugenia Danegger & Sebastián Valverde of a brochure edited in 2017 entitled "*El prejuicio hacia el pueblo Mapuche como supuestamente "chileno": una invención sumamente interesada y largamente refutada*" (Maragliano et al. 2017). http://antropologia.institutos.filo.uba.ar/sites/antropologia.institutos.filo.uba.ar/files/PDTS-CIN-CONICET-UBA-UNCO-%20Folleto%20refutando%20prejuicios%20hacia%20el%20pueblo%20MAPUCE.pdf.

[29]The Mapuche ("people of the land" in Mapudungun) are one of the indigenous groups living in the current Argentine State, although the vast majority reside in the neighboring country of Chile. Based on data from 2018 (Instituto Nacional de Estadísticas 2018), Chile has a total of 17,000,000 inhabitants, of which, 2.100.000 are indigenous people (e.g., Mapuche, Aymara and Diaguita). There are just over 1,700,000 Mapuche. They are settled in the South of Chile, in the Eighth, Ninth and Tenth regions, as well as in the Metropolitan region (around the capital, in the center of the country) due to internal migration processes. In Argentina, the Mapuche are also the most numerous indigenous groups. Of a population of 45,000,000, there are just over 200,000 Mapuche. Nevertheless, Mapuche people in Argentina are geographically more dispersed than other indigenous groups that live in the North of Argentina—mainly qom, kollas, diaguitas, quechua and wichi—(INDEC 2012). The Mapuche are mainly settled in Patagonia, the southernmost region of South America (which includes the provinces of Chubut, Neuquén, Río Negro, La Pampa, Santa Cruz and Tierra del Fuego). As a result of migrations, they also live in the province of Buenos Aires (Radovich and Balazote 2009) and the Metropolitan Area of Buenos Aires.

Argentines from foreigners. The strategy of the political elites consisted in building an "original" link between the territories to be conquered and the original peoples. In this scenario, the Mapuche were identified as "beyond the borders".

In the last decades of the XX century, this fallacious statement has been retaken by diverse social actors, such as journalists and politicians. In order to achieve certain goals, they base their claim in the publications of some authors,[30] whose research has already been refuted by the current academic consensus.[31] Mapuche organizations, diverse academic fields and various institutions have collaborated to refute the fallacies of such explanations in scientific terms. Current research has demonstrated that such discourses, far from being "neutral" and "objective", are closely linked to the interests of different power actors.[32]

Unfortunately, despite the important achievements in Argentina regarding the social and legal recognition of indigenous peoples, these representations are widely spread in the collective imagination. They work as a "historical truth". We find that the persistence and even the exacerbation of the mentioned prejudices are linked to current conflictive situations that are related to the economic interests of different private enterprises on the territories in which the Mapuche people are settled.[33]

Indigenous peoples in Argentina have been historically hidden and invisibilized from local narratives. These discourses have usually highlighted "white pioneers", usually "European"—to a lesser extent also Asian or North-American—, as the main characters of the region's "progress" and the establishment of an integrationist homogenization of culture(s) (Bartolomé 2003a). We are not denying or questioning the significance of the role that inhabitants/settlers from different backgrounds have had in the development and progress of the region. However, this chapter looks to highlight the relevance of the important contributions of the indigenous peoples in the region.

This chapter seeks to contribute, through this collaborative effort between Mapuche communities and organizations, social organizations, several institutions, universities and scientific circles, to problematize the fallacious representations about the Mapuche people. In order to achieve this goal, we propose to examine the diverse fallacious statements and prejudices, through contextualization and historization. Our aim is to provide historical and empirical findings based on our

---

[30]Most of the articles usually base their statements on the work of Casamiquela (1965, 2007).

[31]Examples of authors that have debunked these theories are Radovich and Balazote (2009), Briones (2007), Bartolomé (2003a), Balazote and Valverde (2017), Tozzini and Sabatella (2020), Muzzopappa and Ramos (2017), Collinao et al. (2019), Nagy (2019), de Jong (2018), Trinchero et al. (2018), Kropff (2018), Delrio (2017), Lenton (2017), Maragliano et al. (2017), Vezub (2016), Radovich et al. (2014), Salomón Tarquini (2010), Trentini et al. (2010). For more authors that discuss this topic see Zapata (2010).

[32]The research of the public service positions in the government of the authors usually quoted in order to justify such discriminatory assertions towards the Mapuche people showed that most of them have played important roles in some key periods, such as civic-military dictatorships.

[33]Stigmatizing and criminalizing discourses have shown to be deepened in recent conflicts between private actors, the Chilean and Argentine States and the Mapuche (Balazote and Valverde 2017; Baines et al. 2017).

extensive research looking to refute the biased representations as well as to make the presence of the Mapuche people visible in the Nahuel Huapi[34] Lake.

First, we introduce four of the most common fallacious statements and provide diverse researched elements that refute these fallacies. We prove with overwhelming historical evidence the early presence of Mapuche inhabitants in the current Argentine territory and review the reasons why some Mapuche people born in the current Argentine territory were listed as "Chilean". Finally, we examine how the historical prejudices are being activated over again in current territorial conflicts.

Fallacious Statement 1: The Dichotomy Between "Chilean Mapuche" and "Argentinian Tehuelche"

In the Argentinian traditional historical narrative, a fallacious dichotomy has been imposed and crystallized that holds that whereas the Tehuelche people are truly Argentines, the Mapuche are foreigners that are originally from Chile. This construction omits the fact that these indigenous groups moved freely around the region (Radovich et al. 2014; Balazote et al. 2014; Bellelli et al. 2008; Berón 2007; Mera and Munita 2008) even before the geopolitical boundaries were settled and Argentina or Chile were "invented" as Nation-States. As stressed, the Andes Mountains became an effective border between Argentina and Chile only after the establishment of both republics. Current historical studies demonstrated that the geopolitical limit between the two republics was activated in the 1930s because of economic changes linked to protectionist measures and the reaffirmation of State organizations (Bandieri 2009; Méndez 2006). During that period, the presence of the Argentine State in the region was consolidated through diverse institutions. Additionally, intercommunications between the towns of Argentina's hinterlands and its capital Buenos Aires were improved. For example, the Ferrocarril del Sud to Bariloche started to arrive in 1934 (Bandieri 2009; Méndez 2006). Therefore, the forced and erroneous attribution of an "Argentine" or "Chilean" nationality to indigenous peoples is completely incoherent.

On the other hand, the problem of this stigmatizing dichotomy is that it overlaps more misconceptions. First of all, indigenous groups are mostly depicted as homogeneous entities, and the differences between one another are being exaggerated.[35] The referred traditional postures also avoid contextualizing the history of

---

[34]The spelling in Mapudungun would be Nawel Wapi. It loosely translates into Tiger Island, although this is an inaccurate translation because Spaniards used to refer to Jaguars as Tigers, so a more precise translation would be Jaguar Island.

[35]Examples of this are the emblematic column written by Rolando Hanglin (2009), which had enormous media repercussion or the book by Porcel (2007), as well as a large amount of journalistic articles, many of them published anonymously, given that they are written by authors that are not specialized in the subject. An example of this kind of articles is *Los últimos Tehuelches* [The last Tehuelche] published in the "La Angostura Digital" newspaper (Anonymous 2008).

the different indigenous peoples and partialities. By doing this, the complex and long processes and transformations developed through the centuries, as well as the specific group articulations and complementarities between Mapuche and Tehuelches and/or with the Spanish-Creol society in both sides of the Andes, are being denied.[36] Not acknowledging this process interprets the interaction between the two analyzed groups in simplistic terms. The transformation of the different indigenous groups is consequently and wrongly understood as "extinction", "disappearance", or in terms of "invasion" or "absorption".

Fallacious Statement 2: The "Loss" of Culture

One central argument that is linked to the supposed "foreignness" of the Mapuche people consists of challenging its "indigenous" character. This draws upon the argument of an idealized "racial purity" and a conjectured present-day "loss" of it, which have been refuted in the different academic fields. Although it has been widely demonstrated that it is impossible to establish an identity based on just one cultural trait, especially when indigenous peoples in Latin America were forced to abandon their own language and adopt the language of the Spanish and/or Portuguese colonizer, obsolete criteriums of the usage of certain features (e.g., the use of a specific language) are usually applied when establishing membership to a given indigenous people.

In contrast to these discourses, there is a wide consensus that has been established in the last decades in the social sciences and the humanities. These essentialist arguments that defined populations as "aggregated features" have been widely refuted. Nowadays, identity is conceived in terms of processes and relationships; in other words, identities are configured in the interrelationship with an "other", an on-going negotiation of boundaries between groups of people (Barth 1976; Bonfil Batalla 1972). Therefore, we argue that the conceptions of identity that malicious discourses and communications that present the identity of the Mapuche people as static or the equivalences such as "a race equals a culture" negate their autonomous development and are fallacious. These stances tend to negate the capacity peoples have to adapt and (re)actualize themselves in their identifications in a dynamic and variable manner (Balazote et al. 2014; Radovich et al. 2014; Trentini et al. 2010).

---

[36]This has been analyzed by authors such as Bengoa 2001; Bello 2014 y 2011; Campos Muñoz (2014) in Chile (See more en Bello, 2011) and in Argentina by Bechis 2005; Mandrini 2007; Mandrini y Ortelli 1995 y 1992; Ortelli 1996; Palermo 1986; Nacuzzi 2005; Nacuzzi y Lucaioli 2014; de Jong 2014; Delrío 2005; Bandieri 2005; Lenton 2000; Quijada 2000; Pichumil y Nagy 2016; Radovich y Balazote 2009; Radovich 2013; Trinchero y Valverde 2014: Valverde 2012; Trentini et al. 2010; Ghioldi 2009; Berón et al. 2009; Méndez 2009; Vezub 2009 y 2005; Briones y Delrio 2007; Berón y Radovich 2007; Crespo y Tozzini 2006; Crespo 2009. See more in: Trinchero et al. 2014 y Zapata 2010.

4. Fallacious Statement 3: The Mapuche People Are "Chilean"

There is overwhelming historical evidence, which arises both from documents and oral accounts, that proves the historical presence of the Mapuche people within the current Argentine State. This knowledge reports the early presence of Mapuche inhabitants in the current Argentine territory, which precedes the historical events of "Conquest of the Desert",[37] the establishment of the Argentine Nation-State, as well as the establishment of the "current border" with the current Chilean State.

To support this statement, we invoke the evidence found in the mountainous region of the Los Lagos Department, in the province of Neuquén.[38]

> "La relación que me hizo el indio Domingo Quintuprai de un viaje emprendido por él en el año 1871, mas o menos, para vender aguardiente a los pehuenches establecidos en la falda oriental de la cordillera, entre los lagos Lacar i Nahuelhuapi,[39] me parece interesante por varias razones" [The account that the indian Domingo Quintuprai gave me of a journey undertaken by him in the year 1871, more or less, to sell spirits to the pehuenches settled on the eastern slope of the mountain range, between the Lacar and Nahuelhuapi lakes, seems interesting to me for various reasons] (Lenz 1895–1897: 4).

The linguist and folklorist Rodolfo Lenz (1895–1897) present the case of Domingo Kintupuray, great-grandfather of the elders of the Lof[40] Kintupuray. He was a Huilliche[41] merchant, who used to travel from one side of the Andes mountain range to the other. This testimony is in complete agreement with oral narratives and documents. We would like to highlight that, in his description, Lenz mentions the Pehuenche people[42] being present between the Lacar and Nahuel Huapi lakes in the year 1871. This proves the presence of Mapuche in this area

---

[37]The military subjugation of the Mapuche people occured at the end of the XIX century. Between the years 1879 and 1885, the government implemented military campaigns in La Pampa and Patagonia, using the euphemism "Conquest of the desert" to refer to it. At the same time, on the western border (currently the state of Chile) a similar military operation took place, referred to as the "Pacification of Araucanía" (Radovich and Balazote 2009). The "Conquest of the Desert" resulted in the extermination and subjugation of thousands of indigenous people. It also resulted in the privatization and concentration of large areas of land, which was necessary for the expansion of the landowner class and the consolidation of the agro-export model. The surviving indigenous population had to resettle on marginal lands, where they developed as a preponderant economic activity the extensive raising of livestock (sheep and goats).

[38]The context presented here is just a small sample of the solid and vast records that are available for review, present in numerous studies, publications and documents.

[39]We are keeping the original spelling, from the end of the XIX century.

[40]In Mapudungun, Lof means a community.

[41]Group that is part of the Mapuche people. In Mapudungun, Huilliche can be translated as "people (che) of the south (huilli)".

[42]Group that is part of the Mapuche people. In Mapudungun, Pehuenche can be translated as "people (che) of the Araucaria (pehuen)". Araucaria or Pewen is a coniferous tree indigenous to the southern Andes.

previous to the so-called Conquest of the Desert and to the establishment of the State border (formal and effective) of both Nation-States.

Another valuable proof is a letter written in the year 1908 by José María Paichil[43] to the Dirección de Tierras y Colonias. In it, Paichil demands the regularization of the lands that had been allotted to him together with Ignacio Antreao in 1902.[44] Among the most revealing pieces of information included in the letter, the following extract is included: "*I have resided in this place for the last twenty years*" (fifth line). This proves that Paichil's settlement in the area dates to the late 1880s. In other words, the referred document proves J.M, Paichil's presence in the region several decades before the establishment of schools, post offices and other institutions by which the Argentine Nation-State made itself present in this area.

"José María Paichil indigena de sesenta años, casado, domiciliado en el lote nueve de la Colonia Nahuel Huapi a Ud. respetuosamente expongo: que hace veinte años resido en este lugar, más tarde llegaron las comisiones de límites..." [José María Paichil, sixty year old indigenous man, married, domiciled in the plot number nine of the Nahuel Huapi colony, I respectfully state to you: that I have resided in this place for the last twenty years, only later did the border commissions arrive..."]. [Excerpt (first four sentences): Letter addressed to the Dirección de Tierras y Colonias del Ministerio de Agricultura de la Nación, written in 1908 on behalf of José María Paichil (he was illiterate), demanding the regularization of the plot allotted to him in 1902 (Record N°118/36, Page 13, Dirección de Parques Nacionales, Archivo Ministerio de Agricultura)].

"...fui vaqueano después que salí de la tribu de mi querido jefe Namuncurá y capitanejo Platero..." [...I was a guide after leaving the tribe of my dear Leader Namunkura, and captain Platero...]. [Excerpt (Line 19 and 20): (Record N°118/36, Page 13 reverse side, Dirección de Parques Nacionales, Archivo Ministerio de Agricultura)].[45]

Another fact that is extremely relevant in this document is that José María Paichil stemmed from the tribe lead by the leader Namunkura,[46] the son of Kallfükura, in the area of Salinas Grandes[47] during the second half of the XIX century. In this letter, Paichil states that he acted as "*guide after leaving the tribe of my dear Leader Namunkura, and captain Platero*".[48] The historical background and stigmatization that we have been referring to of the Mapuche people as

---

[43]José María Paichili established, together with Ignacio Antreao, the Paichil Antrao community.

[44]The plot number of the land was 9 of Colonia Agrícola Pastoril Nahuel Huapi.

[45]See "Lof Paichil Antreao Comunidad mapuche ancestral de la región de Villa la Angostura" (Collinao et al. 2019: 18). http://publicaciones.filo.uba.ar/sites/publicaciones.filo.uba.ar/files/Lof%20Paichil%20Antreao.pdf.

[46]Mapuche names are transcribed following the Mapuzungun orthography, specifically the Unified Alphabet.

[47]Salinas Grandes are currently part of the province of La Pampa.

[48]In the original, "*baqueano después que salí de la tribu de mi querido Jefe Namuncurá, y capitanejo Platero*". For a more detailed approach of this journey and the historical background of the leadership of Kallfükura, Namunkura and captain Platero, see Collinao et al. (2019).

supposedly "Chilean", this document demonstrates that Paichil came, as so many other indigenous families, from the Pampas region of the present-day Argentine territory.[49]

The case of Paichil also shows that the military conquest of the indigenous lands in northern Patagonia gave place to many other instances of expulsion and dispersion of indigenous peoples and led to the search for new areas to settle (Collinao et al. 2019). When in the last months of 1878 the army columns invaded Salinas Grandes, they forced the indigenous inhabitants to flee west. Paichil was probably part of the families that followed captain Platero towards the valleys in the Andes mountain range. Towards the end of the 1880s, Paichil and his families settled on the coast of the Correntoso and Nahuel Huapi lakes, a place where Villa La Angostura is currently found.

Another example is the case of the Kintupuray community, whose presence in the region before the Desert Conquest has long been established. Churruhinca Roux (1993) states:

> "The Quinto (meaning the Kintupuray) were at Correntoso from before Parques was born, in '34. From way before. It was said that they lived there when Roca arrived at Neuquen, in '79 (meaning 1879)" (p. 264).

Fallacious Statement 4: The Interpretation of a "Return" to an Ancestral Territory as a "Chilean Invasion"

One of the most important problems of the prejudices against the Mapuche people analyzed in this text is that they deny the overwhelming evidence of the ancestral presence of the Mapuche population within the current Argentine borders. The contextualization of the historical processes lived by the Mapuche shows that at a certain moment the group was forced to mobilize from these lands. Different sources (Delrío 2005; de Jong 2016; Nagy 2019; Subercaseaux 2103; Navarro Rojas 2008; 2nd army division 1883; Curruhuinca Roux 1993; Briones y Delrio 2002; Maragliano et al. 2017; Balazote et al. 2014; Radovich et al. 2014; Ghioldi 2009; Martinelli 2019; Pérez 2016) stress that the Mapuche were forced to move towards the west of the Andes and to resettle in the current Chilean territory, as the military conquest of the Argentine State at the end of the XIX century was advancing. This fact can also be verified in military reports of the Conquest of the Desert written in 1883:

---

[49]See map of the year 1877 (in Collinao et al. 2019 pages 25 and 26), a year before of the invasion perpetrated by the Argentine Army) showing the territory that corresponds to the "Namunkura Indians", occupying a prominent section of the present-day province of La Pampa. http://publicaciones.filo.uba.ar/sites/publicaciones.filo.uba.ar/files/Lof%20Paichil%20Antreao.pdf.

"Not a single Indian has remained in the territory between the Neuquén and Limay rivers, the Andes Mountains and Nahuel Huapi Lake, they have all been thrown to the west." (2nd army division, 1883: 20).[50]

The quote highlights that the soldiers themself had stated in their conquest reports that the indigenous people have been expelled westwards, to what is nowadays Chile, from the current Argentine territory.

So, what is erroneously interpreted as a "Chilean invasion" implies sometimes a "return" of the subjects to the territory of origin (Maragliano et al. 2017; Trentini et al. 2010). Once the military campaigns ended, as well as many decades later, many of these expelled families or their members decided to "return" to their places of origin. Nevertheless, others decided to remain in the current Chilean territory, maintaining a trans-Andean link.[51] This can be verified, among other ways, through the reiteration of surnames on both sides of the Andes mountain of many of the Mapuche families related to each other.

Fallacious Statement 5: Mapuches Are Foreigners Because They Were Registered as "Chilean" Citizens in Chile

Many of the ancestors of the present-day communities have been registered in Argentina as "Chilean", being this the evidence used in many cases as an argument to demonstrate the alleged "foreignness" by the stigmatizing discourses that try to discredit their legitimate belonging to the Argentine territory.

To explain this phenomenon, it has to be taken into account that the lack of Argentine documentation is due to the absence of the State's Registry Office[52] in the mountain range districts of the present-day Argentine territory, which will carry out massive registrations until well into the XX century. Therefore, many Mapuche

---

[50]In the original: "En el territorio comprendido entre los ríos Neuquén y Limay, Cordillera de los Andes y Lago Nahuel Huapi no ha quedado un solo indio, todos han sido arrojados al occidente". (Segunda división del ejército 1883: 20).

[51]The Andes is the longest continental mountain range in the world, about 7000 km long. It works as a natural geographic border, as well as the legal–political boundary between Chile and Argentina. Although it is a mountain range, it does not constitute an uncrossable barrier, due to the fact that it is also constituted by valleys and paths that are able to be crossed in east–west direction (especially during the summer). The disciplines of Archeology and history have established the articulation of indigenous populations on one side and the other of it through thousands of years.

[52]At the beginning of the XX century, the only registry offices available for the population of the northern Patagonian region were those in San Pablo in Chile and in Villa La Angostura in Argentina, locations which for the locals meant the same region. The registry office was only established in the area in 1897. Therefore, many times people figure as Chilean rather than Argentine.

crossed the Andes to sell their products, buy merchandise and carry out different bureaucratic paperwork[53] and during these travels would write their newborns down in Chile. This explains why many of the newborns in the Argentine territory, which then continued their residency there, were written down as "Chilean" (Sarobe 1935). Given the relations with the departments west of the Andes in the present-day Chilean State, this kept happening well into the XX century.

The Reiteration of Prejudices in Current Territorial Conflicts

As we have already indicated, since the return to democracy in 1983, and especially in recent years, indigenous peoples have been acquiring greater political and legal recognition through the enactment of different laws. These advances have been the result of intense demands made through prolonged struggles and negotiation processes[54] with the different levels and State institutions.[55]

Nevertheless, throughout all of Argentina, like in other countries of the region, there are enormous difficulties to be found for the implementation of the different regulations in force. This happens due to the resistance of various private interests and local power networks, which are at the same time often connected with the State in different ways. These are the sectors that precisely contribute and promote, through stigmatizing discourses, the negative reaction and exclusion of the society towards the Mapuche community.

It should be noted that the different indigenous communities of the country have long been affected by extractivist dynamics in the territories in which they are settled. Extractivism based on agribusiness, the rise in the use of agrochemicals and the "open-pit" or "mega" mining are critically affecting the ecology and environment, the life conditions of the local communities, often leaving epidemiological consequences and sequelae. We must also mention the consequences generated in some contexts by the exploitation of hydrocarbons, the tourism activity or the real estate speculation. This affects, for example, the Mapuche communities of the Nahuel Huapi National Park region and surrounding areas, as well as other touristic areas in the region.[56] All these situations have been getting worse in recent years.

---

[53]J. M. Sarobe (1935) shows this clearly as well as the existing trans-Andean links well into the XX century.
[54]These struggles and negotiation processes have lasted for years and sometimes decades.
[55]See Engelman and Varisco in this book.
[56]See for example Iñigo Carrera, Balazote and Stecher in this book. In it they analyze the case of the Lof Paichil-Antriao, located in the periphery of Villa la Angostura, in the northern area of the Nahuel Huapi national park, in a highly valued zone for the tourism industry. The community has been having conflicts over these valuable lands with several private investors and some federal agents for years. The same could be said of the Lof Quitriqueo, which is settled on the northern shore of the Nahuel Huapi lake, which holds several legal disputes with private agents who want to take over the valuable lands on which they are traditionally settled.

The dynamics that affect Mapuche people in this area regarding the use of their ancestral territories and in relation to their access to forests and other resources must be contextualized. The conflicts are connected to the implementation of Law 26,160 *"Emergency law in the matter of possession and ownership of the lands traditionally occupied by the indigenous communities of the country".*[57] This law has been of high importance for indigenous peoples in Argentina since its sanction in 2006 and its extensions of 2009, 2013 and 2017. By declaring the emergency of the communities' territories, the law has recognized the precarious situation in which these communities are living in terms of formalizing the legal status of their territories, as well as the current precariousness and social vulnerability, among others, that affects them. In addition, the mentioned law is also articulated with other legislation, the National Law 26,331 on Forests. On the tenth point of the Annex, which refers to the "Criteria of environmental sustainability for the territorial zoning of native forests", a framework of application of National Law 26,160 is established. The Forests law states that:

> "In the case of Indigenous Communities and within the framework of Law 26,160, action must be taken in accordance with the provisions of Law 24,071, ratifying Convention 169 of the International Labor Organization (ILO)" (Casalderrey Zapata and Tozzini 2020; Guiñazú 2017).[58]

As long as this regulation is in force, even with the possible critiques regarding its limitations, it constitutes a very helpful tool for the indigenous communities, since it suspends evictions as well as contemplates the implementation of survey projects of the territories and the recognition of the indigenous communities throughout the country. It is a legal instrument designed for the defense of indigenous peoples' territorialities since it grants them a public document (the technical folder) that communities are able to use in different judicial or conflictive instances, as well as to access to other public policies aimed for indigenous communities (Casalderrey Zapata and Tozzini 2020; Guiñazú 2017; Brown et al. 2020).

In fact, the enactment of National Law 26,160 *"on indigenous territorial emergency"* 14 years ago allowed the National Program for Territorial Survey of Indigenous Communities (RETECI)[59] to identify around half of the 1,724 communities in the country.[60] Surveys remain to be carried out in 649 communities, which is equivalent to slightly less than 40% of the total.

Regarding the area in the Bariloche department of the Río Negro province, in which we have been working, all the communities that had legal status at the time

---

[57]In the original: "Ley de emergencia en materia de posesión y propiedad de las tierras que tradicionalmente ocupan las comunidades indígenas originarias del país".
[58]In the original: "En el caso de las Comunidades Indígenas y dentro del marco de la ley 26.160, se deberá actuar de acuerdo a lo establecido en la ley 24.071, ratificatoria del Convenio 169 de la Organización Internacional del Trabajo (OIT)".
[59]In Spanish: "Programa Nacional de Relevamiento Territorial de Comunidades Indígenas (RETECI)". The program depends on the National Institute of Indigenous Affairs (I.N.A.I.).
[60]The survey process has concluded in just over 40% of the surveyed cases. 416 (24%) surveys remain to be completed.

of the surveys have already been registered. Nevertheless, the situation of the Mapuche communities in the Los Lagos department, in the Neuquén province, is quite different, and this leads to the reiteration of prejudices and deployment of violence towards Mapuche people in current territorial conflicts. This happens because the provincial authorities do not legally recognize them, since the implementation of this National Law 26,160 "*of indigenous territorial emergency*" is still pending. Therefore, the application of the Forest Law in the territories of the indigenous communities that would allow them as beneficiaries is still not enabled.[61] This inevitably takes affects the rights of the indigenous groups. For example, the absence of the implementation of the available legal framework took in the last years Lof Paichil Antreao, from the zone of Villa La Angostura to different conflicts with private entrepreneurs.[62] Although the State already recognized the community and its rights over their ancestral territory in 2007, the violation of the national law and the provincial law of Forests has a big impact, not only in environmental and cultural matters but also in increasing criminalizing discourses towards the Mapuche people.

Final Words

In this brief chapter, we tried to problematize some of the most extended fallacies related to the Mapuche people, as they circulate in Argentina. It remains to be clarified that it is very common in times in which the Mapuche people face different conflicts, the reiteration of these fallacious discourses. These imaginaries are usually based on knowledge and publications that are at least 50 or 60 years old, sometimes even older. In many cases, some of the publications are even anonymous, which is intellectually dishonest.

Ultimately, this widespread prejudice is intended to delegitimize the present-day Patagonian indigenous peoples, by presenting them either as "foreign" (the Mapuche) or as "real" and "legitimate" but already extinct (the Tehuelche), that is to say, as part of the past. We find it fundamental to contribute to building a more just and egalitarian society that therefore recognizes its own internal diversity. Because of that extremely entrenched racism that is found in certain segments of Argentine society, a racism that has become exacerbated for example in the context of the COVID-19 pandemic[63], it remains functional to the reproduction of the profound socioeconomic inequality of the present-day Argentine society. To keep up the exploitation, to keep indigenous peoples occupying the most precarious and worst paid position in the production structure, some segments of the society find it

---

[61]See Iñigo Carrera, Balazote and Stecher in this volume.
[62]See for example: (Roncarolo 2020) https://www.rionegro.com.ar/la-corte-interamericana-de-derechos-humanos-interviene-en-un-conflicto-mapuche-1302750/.
[63]See "Notes on the academic report on the impact of COVID-19 on indigenous peoples" in this book.

necessary to degrade the "other", depriving them of their rights (Tamagno 2019). We believe it is necessary to reassert, as Miguel Bartolomé has pointed out, that the indigenous peoples:

"(…) are not a testimony of the past or an anachronism, as some nationalist perspectives anchored in nineteenth-century ideologies still claim, but a vibrant part of the present and a desirable part of the future. And of a future that, to avoid repeating historical injustices, it should be erected on the acceptance and mutual respect between the multiple and different ways of being a member of a State formation" (Bartolomé 2003b: 201).[64]

References

Anonymous (2008) Los últimos Tehuelches (Part I, October 20th, 2008 and Part II, October 21, 2008). La Angostura Digital. Last access: 9 Oct 2020. http://www.laangosturadigital.com.ar/v3.1/home/interna.php?id_not=6935&ori=web

Baines SG, Balazote A, Berón M, Castilla M, Engelman J, Lustosa IMC, Molek N, Pérez A, Quiroga L, Radovich JC, Trinchero H, Valverde S, Weiss ML (2017) A ABA e seu Comitê Povos Tradicionais, Meio Ambiente e Grandes Projetos subscreve o documento abaixo que denuncia e confronta a campanha de estigmatização e criminalização dos Povos Indígenas na Argentina. Associação Brasileira de Antropologia. Last access: 9 Oct 2020. http://www.aba.abant.org.br/files/20170913_59b9836213ffa.pdf

Balazote A, Radovich JC, Berón M, Valverde S, Stecher G (2014) Deslegitimación y discriminación en el discurso mediático. Agencia Ciencia, Tecnología y Sociedad. Instituto de Medios de Comunicación, Universidad Nacional de La Matanza, La Matanza. Last access: 9 Oct 2020. http://www.ctys.com.ar/index.php?idPage=20&idArticulo=3028

Balazote A, Valverde S (20 de septiembre 2017) Contra la criminalización del pueblo mapuche. Infonews. Last access: 9 Oct 2020. http://www.infonews.com/

Bandieri S (2009) Cuando crear una identidad nacional en los territorios patagónicos fue prioritario. Revista Pilquen 11(1):1–5

Barth F (1976) Introducción. In Barth F (ed) Los grupos étnicos y sus fronteras. La organización social de las diferencias culturales. Fondo de Cultura Económica, Ciudad de México, pp 9–49

Bartolomé MA (2003a) Los pobladores del 'Desierto' genocidio, etnocidio y etnogénesis en la Argentina. Cuadernos de Antropología Social 17:162–189

Bartolomé MA (2003b) En defensa de la etnografía. El papel contemporáneo de la investigación intercultural. Revista de Antropología Social 12:199–222

Bellelli C, Scheinsohn V, Podestá M (2008) Arqueología de pasos cordilleranos: un caso de estudio en Patagonia norte durante el holoceno tardío. Boletín del museo chileno de arte precolombino 2(13):37–55

Bello Á (2011) Nampülkafe. El viaje de los mapuches de la Araucanía a las pampas argentinas. Territorio, política y cultura en los siglos XIX y XX. Ediciones Universidad Católica de Temuco, Temuco

Berón M (2007) Circulación de bienes como indicador de interacción entre las poblaciones de la pampa occidental y sus vecinos. Arqueología en las Pampas. In: Bayón C, Pupio A, González MI, Flegenheimer N, Frére M (eds) Sociedad Argentina de Antropología, Buenos Aires, pp 345–364

[64]In the original: "No son un testimonio del pasado o un anacronismo, como todavía lo pretenden algunas perspectivas nacionalistas ancladas en las ideologías decimonónicas, sino parte integrante y viva del presente y deseablemente del futuro. Y de un futuro que, para evitar reiterar injusticias históricas, deberá ser construido a partir de la aceptación y el respeto entre las múltiples y diferentes formas de ser miembro de una formación estatal".

Bonfil Batalla G (1972) El concepto de indio en América: una categoría de la situación colonial (1972) Anales de Antropología. Revista del Instituto de Investigaciones Antropológicas, vol 9. UNAM

Briones C (2007) Nuestra lucha recién comienza. Vivencias de pertenencia y formaciones mapuche de sí mismo. Ava 10:23–46

Brown A, Castelnuovo N, Castilla M, Engelman J, Guiñazu S, Valverde S (2020) Anexo XLIII: Ley N.° 26.160 y la emergencia territorial indígena. In: Abeledo S et al (eds) Informe ampliado: efectos socioeconómicos y culturales de la pandemia COVID-19 y del aislamiento social, preventivo y obligatorio en los Pueblos Indígenas en Argentina. Segunda etapa, junio 2020, pp 314–320. Last access: 9 Oct 2020. http://antropologia.institutos.filo.uba.ar/sites/antropologia.institutos.filo.uba.ar/files/info_covid_2daEtapa.pdf

Casalderrey Zapata C, Tozzini MA (2020) De contar árboles a pintar su ubicación. Gubernamentalidad y aplicación de la Ley 26331 de Bosques Nativos en Patagonia 1. Tabula Rasa 34:131–153. https://doi.org/10.25058/20112742.n34.07

Casamiquela R (1965) Rectificaciones y ratificaciones hacia una interpretación definitiva del panorama etnológico de la Patagonia y área septentrional adyacente. Colección 'Instituto de Humanidades Cuadernos del Sur'. Antropología cultural y social. Patagonia. Universidad Nacional del Sur (Argentina). Instituto de Humanidades, Bahía Blanca

Casamiquela R (2007) Rodolfo Casamiquela racista antimapuche: o la verdadera antigüedad de los mapuches en la Argentina. Edited by the author, Buenos Aires

Churruhinca Roux C (1993) Las matanzas del Neuquén. Crónicas Mapuces. Editorial Plus Ultra, Buenos Aires

Collinao F, Loncón L, Olivero D, Subiri L, Tropan S, Márquez V, Florentino N, Ghioldi G, Trinchero H, Balazote A, Radovich JC, Ramos M, de Jong I, Maragliano G, Impemba M, Stecher G, Valverde S, Varisco S, Pérez A (2019) Lof Paichil Antreao: comunidad mapuche ancestral de la región de Villa la Angostura. Editorial de la Facultad de Filosofía y Letras, Universidad de Buenos Aires, Ciudad Autónoma de Buenos Aires

De Jong I (2016) Las prácticas diplomáticas en los procesos de expansión estatal: Tratados de Paz y Parlamentos en Pampas y Araucanía (1850–1880). In: de Jong I, Escobar Ohmstede A (eds) Las poblaciones indígenas en la construcción y conformación de las naciones y los estados en la América Latina decimonónica. CIESAS, El Colegio de México y El Colegio de Michoacán, Ciudad de México, pp 291–348

De Jong I (2018) Guerra, genocidio y resistencia: apuntes para discutir el fin de las fronteras en Pampa y Norpatagonia, siglo XIX. Revista Habitus 16(2):229–254

Delrio W (2005) Memorias de expropiación. Sometimiento e incorporación indígena en la Patagonia 1872–1943. Buenos Aires, Universidad Nacional de Quilmes

Delrio W (2017) La lucha de los mapuches y sus estereotipos. Nueva sociedad

Ghioldi G (Comp) (2009) Historia de las familias mapuche Lof Paichil Antriao y Lof Quintriqueo: mapuche de la margen norte del lago Nahuel Huapi. Archivos del Sur, Biblioteca Popular Osvaldo Bayer, Neuquén

Guiñazú S (2017) La performatividad de las políticas públicas: modalidades de interacción e interpelación entre Estado, sociedad e indígenas en el proceso de ejecución de una política pública indigenista, 2006–2017. Revista Estado y Políticas Públicas 5(9):145–167

Hanglin R (22 de septiembre 2009) La cuestión mapuche. La Nación, Sección Opinión

INDEC (2012) Censo Nacional de Población, Hogares y Viviendas 2010 - Censo del Bicentenario: resultados definitivos, Serie B. Número 2. Buenos Aires: Instituto Nacional de Estadística y Censos (INDEC)

Kropff L (2018) Emoción e identidad en relatos biográficos de jóvenes mapuche a principios del siglo XXI. Revista de antropología social y cultural Etnografías contemporáneas 4(7):83–109

Lenton D (2017) El nuevo enemigo público. La criminalización de los mapuches. Revista Anfibia. Last access: 15 Oct 2020. http://www.revistaanfibia.com

Lenz R (1895–1897) Materiales para el estudio de la lengua, la literatura y las costumbres de los indios mapuche o araucanos. Universidad de Chile, Santiago de Chile

Maragliano G, Impemba M, Stecher G, Valverde S, Balazote A, Berón M, Pérez A, Radovich JC (2017) El prejuicio hacia el pueblo Mapuce como supuestamente "chileno": una invención sumamente interesada y largamente refutada. In: Poblaciones Mapuce de la cuenca del Nahuel Huapi: hacia el reconocimiento identitario y sus derechos socioeconómicos y culturales. Proyecto PDTS-CIN-CONICET, Ciudad de Buenos Aires

Martinelli ML (2019) Procesos de territorialización y reservas indígenas en Ñorquin-có: apuntes para la reconstrucción de las trayectorias de las comunidades de Ancalao y Cañumil (1900–1950). In: Cañuqueo L, Kropff L, Pérez J y Wallace (eds) La tierra de los otros. Editorial de la Universidad Nacional de Río Negro, Bariloche, pp 67–94

Méndez L (2006) Circuitos económicos en el Gran Lago. La región del Nahuel Huapi entre 1880 y 1930. In: Bandieri S, Blanco G, Varela G (eds) Hecho en Patagonia. La historia en perspectiva regional. Ediciones Centro de Estudios Históricos Regionales, Universidad del Comahue, Neuquén

Mera R, Munita D (2008) Informe ejecutivo salvataje sitio "Villa JMC-01-Labranza". Comuna de Temuco, Provincia de Cautín, región de la Araucanía, p 57

Muzzopappa ME, Ramos AM (2017) Encontrar al terrorista. De la seguridad nacional al código penal (en)clave Comahue. Revista patagónica de estudios sociales 22:101–120

Nagy M (2019) Genocidio: derrotero e historia de un concepto y sus discusiones. Memoria americana. Cuadernos de etnohistoria 2(27):10–33

Navarro Rojas L ([1909] 2008) Crónica militar de la conquista y pacificación de la Aracuanía desde el año 1859 hasta su completa incorporación al territorio nacional. Editorial Pehuén, Santiago de Chile

Pérez P (2016) Archivos del silencio. Estado, indígenas y violencia en Patagonia Central, 1878–1941. Prometeo, Buenos Aires

Porcel RE (2007) La araucanización de nuestra pampa. Los tehuelches y pehuenches. Los mapuches invasores. Edited by the author, Buenos Aires

Radovich JC, Balazote A (2009) El pueblo mapuche contra la discriminación y el etnocidio. In: Ghioldi G (Comp) Historia de las familias mapuche Lof Paichil Antriao y Lof Quintriqueo: mapuche de la margen norte del lago Nahuel Huapi. Neuquén: Archivos del Sur, Biblioteca Popular Osvaldo Bayer, Ferreyra editor, pp 35–59

Radovich J, Balazote A, Berón M, Valverde S, Stecher G (25 de octubre 2014). Los argumentos falaces sobre el pueblo mapuche. Página 12. Last access: 9 Oct 2020. https://www.pagina12.com.ar/diario/sociedad/3-258339-2014-10-25.html

Roncarolo L (27 de marzo 2020) La Corte Interamericana de Derechos Humanos interviene en un conflicto mapuche. Diario Río Negro. Last access: 9 Oct 2020. https://www.rionegro.com.ar/la-corte-interamericana-de-derechos-humanos-interviene-en-un-conflicto-mapuche-1302750/

Salomón Tarquini C (2010) Largas noches en La Pampa. Itinerarios y resistencias de la población indígena (1878–1976). Prometeo, Buenos Aires

Sarobe JM (1935) La Patagonia y sus Problemas. Estudio Geográfico, Económico, Político y Social de los Territorios Nacionales del Sur. Aniceto López, Buenos Aires

Segunda División del Ejército (1883) Campaña de los Andes. Al sur de la Patagonia. Partes detallados. Diario de la Expedición. Buenos Aires

Subercaseaux F ([1833] 2013) Memorias de la Campaña a Villarrica 1882–1883. In Villalobos S (Coord) Incorporación de la Araucanía. Relatos militares 1822–1883. Catalonia, Santiago, pp 285–351

Tamagno L (2019) Reflexiones sobre el malestar. Pensando la etnografía. In Tola F et al. (eds) Malestar en la etnografía. Malestar en la antropología. Instituto de Desarrollo Económico y Social, Ciudad Autónoma de Buenos Aires, pp 48–68

Tozzini MA, Sabatella ME (2020) Es un ídolo con pies de sal a punto de vadear un río. Apreciaciones sobre el relevamiento territorial de la ley 26,160 en dos causas judiciales de comunidades mapuche en el Maitén, provincia de Chubut. Revista Papeles de Trabajo 23:13–29

Trentini F, Valverde S, Radovich JC, Berón M, Balazote A (2010) Los nostálgicos del desierto: la cuestión mapuche en Argentina y el estigma en los medios. Cultura y representaciones sociales 8(4)

Trinchero H, Balazote A, Radovich JC, Castilla M, Engelman J, Valverde S (2018) Pueblos indígenas en Argentina: fronteras históricas y contemporáneas. In de Brum AK, Espósito Neto T, Contini AAM (Orgs) Desenvolvimento para além das Fronteiras: diálogos sobre aspectos sociais, culturais y regionais Curitiba: Appris, pp 91–126

Unidad de Estudios y Estadísticas de Género (2018) Radiografía de Género: Pueblos Originarios en Chile 2017. Instituto Nacional de Estadísticas, Chile. Last access: 28 Sept 2020. https:// historico-amu.ine.cl/genero/files/estadisticas/pdf/documentos/radiografia-de-genero-pueblos-originarios-chile2017.pdf

Vezub JE (2016) El estado sin estado entre los araucanos/mapuches chungara. Revista de Antropología chilena 48:723–727

Zapata HMH (2010) Pensar El Bicentenario Argentino desde y con los Pueblos Indígenas: descolonizando memorias, identidades y narrativas. Revista Mosaico-Revista de Historia 3(2)

Printed in the United States
by Baker & Taylor Publisher Services